Lecture Notes in Mathematics

Edited by A. Dold and B. Eckmann

Series: Mathematisches Institut der Universität Bonn
Adviser: F. Hirzebruch

815

Peter Slodowy

T0222458

Simple Singularities and Simple Algebraic Groups

Springer-Verlag
Berlin Heidelberg New York 1980

Author

Peter Slodowy
Mathematisches Institut der Universität Bonn
Wegelerstr. 10
5300 Bonn
Federal Republic of Germany

AMS Subject Classifications (1980): 14B05, 14D15, 17B20, 20G15

ISBN 3-540-10026-1 Springer-Verlag Berlin Heidelberg New York
ISBN 0-387-10026-1 Springer-Verlag New York Heidelberg Berlin

Printed in Germany

Printing and binding: Beltz Offsetdruck, Hemsbach/Bergstr.
2141/3140-543210

Introduction

By a rational double point or a simple singularity (in the introduction, say over $\mathbb{C}$, for simplicity) we understand the singularity of the quotient of $\mathbb{C}^2$ by the action of a finite subgroup of $SL_2(\mathbb{C})$. In the minimal resolution of such a singularity an intersection configuration of the components of the exceptional divisor appears which can be described in a simple way by a Dynkin diagram of type A_r, D_r, E_6, E_7 or E_8. Up to analytic isomorphism, these diagrams classify the corresponding singularities (for details see 6.1). Moreover, these diagrams also classify just those simple Lie algebras and Lie groups which have root systems with only roots of equal length.

Besides the connection between the rational double points and Dynkin diagrams mentioned above, which has been known since the work of Du Val (cf. [DV]) and goes back essentially to the claim that the integral intersection form for the components of the exceptional divisor is negative definite (cf. [M1], [Ar1]), further connections were discovered by Brieskorn in the works [Br1] and [Br3] in investigating the simultaneous resolution of holomorphic maps with simple singularities, connections with various other structures linked to Dynkin diagrams such as the Weyl groups, Weyl chambers, the Coxeter numbers and the Coxeter transformations.

With knowledge of these results and those of Kostant and Steinberg (cf. [Ko2], [St1]) on the quotients $\gamma : \underline{g} \to \underline{h}/W$ and $\chi : G \to T/W$ of a simple Lie algebra $\underline{g}$ and Lie group G by the operation of the adjoint group, as well as the results of Springer and Tits on the resolution of the variety of nilpotent resp. unipotent elements in $\underline{g}$ resp. G (cf. [Sp2] 1.4, 1.5), Grothendieck conjectured the following connections between the rational double points and the simple Lie algebras (and analogously, the simple Lie groups) of type A_r, D_r, E_r:

i) The intersection of the variety $N(\underline{g})$ of the nilpotent elements of $\underline{g}$ with a transverse slice S to the so-called subregular orbit, which has codimension 2 in $N(\underline{g})$, is a surface $S \cap N(\underline{g})$ with an isolated rational double point of the type corresponding to the algebra $\underline{g}$.

ii) The restriction of the quotient $\gamma : \underline{g} \to \underline{h}/W$ to the slice S is a realization of a semiuniversal deformation of the singularity in $S \cap N(\underline{g})$.

Moreover, Grothendieck arrived at a generalization of the cited results of Springer ([Sp2] 1.4) in the simultaneous resolution of the quotient $\gamma : \underline{g} \to \underline{h}/W$ (resp. $\chi : G \to T/W$) which, if the conjectures above were true, would induce the simultaneous resolutions explicitly constructed by Brieskorn ([Br1], [Br3]).

In fact, Grothendieck's conjectures were then proved by Brieskorn. A short suggestion for a proof is found in the Reports of the International Mathematical Congress 1970 in Nice ([Br4]). Although Brieskorn's proof is logically independent of the resolution of the nilpotent variety (cf. 8.3), the results of Tits on Dynkin curves, i.e. on the structure of the fibers of the resolution over the subregular elements (cf. [St2] p. 148), played a definite rôle, since except for knowledge of the self-intersection number they gave exactly the exceptional divisor of the minimal resolution of the singularity in question.

Not until recently, in the thesis of Hélène Esnault ([Es]) written under the guidance of Lê Dũng Tràng, was the calculation of the self-intersection numbers for the components of the exceptional divisor in the resolution of the surface $S \cap N(\underline{g})$ finally accomplished. In this way conjecture i) was proved by using the characterization of the rational double points by the structure of their minimal resolutions.

Except for some indications no derivation for most of the results mentioned above can be found in the literature. One goal of this work is to develop detailed proofs of the results in question within the more general framework of algebraic geometry over algebraically closed fields (with some slight restrictions on the characteristic). Another goal is the extension of these results to also include the simple Lie algebras and Lie groups with inhomogeneous root systems, which are classified by the diagrams B_r, C_r, F_4 and G_2. In these cases, the singularity of the intersection of the nilpotent variety $N(\underline{g})$ (or the unipotent $V(G)$) with a transverse slice S to the subregular orbit can be identified as a singularity of type A_{2r-1}, D_{r+1}, E_6 and D_4. Now the question is, which deformation of these singularities is realized by the restriction of the quotient $\gamma : \underline{g} \to \underline{h}/W$ to the slice S. The diagrams B_r, C_r, F_4, G_2 can be interpreted in a symbolic way as quotients of the homogeneous diagrams A_{2r-1}, D_{r+1}, E_6, D_4 by the operation of certain diagram symmetries. Moreover, these diagram symmetries can be realized in two natural, essentially equivalent ways: on one hand in the Lie algebraic context by the action of the centralizer of a subregular element on the components of the corresponding Dynkin curve, on the other hand a priori, i.e. without any reference to Lie groups, by the action of the quotient F'/F of certain finite subgroups of SL_2 on the quotient singularity $\mathbb{C}^2/F$ and its minimal resolution. The deformations in the Lie algebras with inhomogeneous root system are not in fact semiuniversal in the usual sense but rather are semiuniversal with respect to those deformations all of whose fibers are operated on by the same relevant group of diagram symmetries.

As an essential aid to our investigation we will use the theory of deformations of isolated singularities with group actions, which we develop in Part I for certain complete intersections with linearly reductive group actions.

In Part II we will first summarize (in chapter 3), with several small technical additions, the known results about the quotient morphisms $\chi : G \to T/W$ and $\gamma : \underline{g} \to \underline{h}/W$ and then obtain (Chapter 4) the simultaneous resolution of these morphisms as a special case of a more general construction which gives similar results for the closures of the so-called "Dixmier sheets".

In Part III, after some technical preparations concerning transverse slices (Chapter 5), we identify the singularities of the nilpotent resp. unipotent variety along its subregular orbit (Chapter 6). For that we use the results about the structure of the Dynkin curves as derived in detail in [St2], and also reproduce the calculation of the self-intersection numbers, somewhat modified from that of [Es].

Part IV will finally deal with deformations of simple singularities as realized in the simple Lie algebras and Lie groups and proves their semiuniversality (8.7). As in the sketch given by Brieskorn in [Br4], the quasihomogeneous structure of the semiuniversal deformation of a simple singularity plays an essential rôle. Besides the theory of deformations with group actions, some consequences of the Jacobson-Morozov-Lemma for the structure of nilpotent elements in simple Lie algebras will be needed here (7.1 - 7.4). Moreover, the proof of the analogue to Grothendieck's conjecture (ii) for the cases B_r, C_r, F_4 and G_2 demands a detailed study of the so-called reductive centralizers of subregular elements and their actions on the Lie algebra and the simple singularity that are concerned (6.2, 7.5, 7.6, 8.4, 8.8). As an application we determine, among other things, the configurations and types of the singularities of the neighboring fibers of semiuniversal deformations of simple singularities (both with and without symmetries).

The correspondence between simple singularities and simple Lie groups also makes sense over non-algebraically-closed fields, where "forms" of singularities and groups have to be considered. In Appendix I we state the main results leaving details and elaborations to a later work. Appendix II shows that the adjoint quotients $\gamma : \underline{g} \to \underline{h}/W$ themselves may be regarded as semiuniversal deformations in an appropriate sense. Finally Appendix III extends an observation of John McKay concerning the representation theory of binary polyhedral groups and homogeneous affine Dynkin diagrams to the "relative" representation theory of these groups and inhomogeneous affine Dynkin diagrams.

In this work we have not dealt with the aspect of monodromy of simple singularities. The interested reader may look into [Sl 2] where in a more general situation monodromy representations of Weyl groups are introduced which are related to a recent construction of Weyl group representations by T. A. Springer.

This work is a revised, enlarged and translated version of the former work "Einfache Singularitäten und einfache algebraische Gruppen" which appeared as Regensburger Mathematische Schrift 2 (1978) and which by now is out of print.

Besides a number of simplifications, corrections, additional remarks and three appendices there is one important novelty. Whereas in the first version the nature of the symmetries on the simple singularities which describe the subregular deformations in Lie groups of type B_r, C_r, F_4, G_2 remained mysterious, we now give an a priori definition for them in terms of binary polyhedral groups leading to the notion of a simple singularity of type B_r, C_r, F_4, G_2. We also give a geometric characterization of these symmetries (new section 6.2). Accordingly the central parts of chapter 8 have been reorganized and rewritten. Also, for the convenience of the reader, we have rewritten chapter 2 (less technically) and added an example (new section 3.11).

The translation of the basic text was done by Daniel P. Johnson, Madison Wisconsin, whom I wish to thank here for his efforts in completing this task. The final redaction was supported by the Sonderforschungsbereich "Theoretische Mathematik", Universität Bonn.

Besides many others who have contributed to this work through their suggestions and discussions, I especially want to thank E. Brieskorn, Th. Bröcker, H. Esnault, H. Kraft and H. Pinkham.

Bonn, March 1980

Peter Slodowy

Table of Contents

IV Deformations of Simple Singularities

7. Nilpotent Elements in Simple Lie Algebras

8. Deformations of Simple Singularities

Hints for the Reader

In this presentation, we do not steer directly for the goals mentioned in the introduction. For example, in Part II we have given more attention to restrictions on the characteristic of the base field than is necessary for later applications. Moreover we have gone into several particulars of the situation for groups which are not of interest for the questions in deformation theory. For those readers who wish a fast passage to the proof of the main result (Theorem 8.7) for the types A_r, D_r, E_6, E_7, E_8 with as little use as possible of the results about algebraic groups, the following sections are suggested: Part I (first used in Chapter 8), 3.1 - 3.4, 3.8 - 3.10, 3.12, 3.14, 5.1, 5.2, 5.4, 5.5, 6.1, 6.5 (after 8.3), 7.1 - 7.4, 8.1 - 8.3, 8.6, 8.7. To aid simplicity, one can also assume that $k = \mathbb{C}$ so that questions of separability (for $\mathrm{char}(k) > 0$) or êtale morphisms (locally analytic isomorphisms) can be passed over. The reader may also profit from reading the surveys given in [Br 4] and [Sl 2]. Except for certain exceptions (localization, completion) we work in the framework of separated algebraic schemes over an (except for Part I and Appendix I) algebraically closed field k (EGA I 6.5). We identify such schemes, when they are reduced, with their k-rational points, as is usual in group theory ([Bo]).

I. Deformations with Group Actions

1. Regular Group Actions

1.1. Conventions. In part I, i.e. chapters 1 and 2, the following conventions will be valid. By k we denote a commutative field. All (formal) k-schemes considered will be affine and Noetherian unless otherwise stated. By a k-variety we mean a k-scheme of finite type possibly not reduced (algebraic in the sense of EGA I 6.5). All algebraic groups will be affine and defined over k in the sense of [Bo], that is reduced over an algebraic closure $\bar{k}$ of k. By $\mathbb{A}^n_k$ we denote the spectrum $\operatorname{Spec} k[x_1,\dots,x_n]$ of the polynomial ring in n indeterminates.

1.2. Actions. An action of a k-group on a (not necessarily affine) k-scheme, defined by a k-morphism $G \times X \to X$, is called regular. Every formal scheme X can be regarded as an inductive limit $X = \varinjlim X_i$ of schemes (EGA I 10.6). A regular action of G on X will thus be the inductive limit of regular actions $G \times X_i \to X_i$.

If G operates linearly on a k-vector space V, then it also operates in a natural way on the dual V^* (contragradiently), and on all associated tensor products. Of particular interest is the fact that G operates linearly on the symmetric products (powers) $S^n(V) = S_n(V^*)$ of V^*, and on their direct sum $\bigoplus_{n=0}^{\infty} S^n(V) = S^*(V)$, which can be viewed as the affine algebra $k[V]$ of the variety V.

In general, with the action of G on a k-scheme X, we get a linear action on the global sections $k[X]$, and in the event that x is a fixed point of G in X, we also get one on the local ring of X in x and its completion.

1.3. Linearization. Let the k-group G operate regularly on an affine k-scheme $X = \operatorname{Spec} R$. Then G operates linearly on R. A lemma of Cartier states that every $f \in R$ lies in a finite-dimensional G-invariant k-vector subspace of R (see [Bo] 1.9 or [M2] § 1, 13). With that, we can easily show the following:

Proposition: Let the k-group G operate regularly on the k-variety X. Then there are regular linear actions of G on finite-dimensional k-vector spaces V and W, a G-equivariant k-morphism $\phi : V \to W$ and a G-equivariant k-isomorphism $X \tilde{\to} \phi^{-1}(0)$.

Proof: The affine ring $R = k[X]$ of X is by assumption a finitely generated k-algebra. By Cartier's lemma, there is, therefore, a finite-dimensional G-invariant k-vector subspace V^* of R, which generates R. The inclusion $V^* \hookrightarrow R$ induces a G-equivariant surjective homomorphism $k[V] = S_*(V^*) \overset{\pi}{\to} k[X]$, which gives an equivariant embedding $X \hookrightarrow V$.

The kernel J of π is a G-submodule of $k[V]$, and is also a finitely generated ideal. As in the first part, we can find a finite-dimensional G-equivariant subspace $W^* \subset J$ which generates J as a $k[V]$-ideal. The inclusion $W^* \hookrightarrow k[V]$ induces a G-equivariant homomorphism $k[W] = S_*(W^*) \to J \subset k[V]$, which defines the morphism we are seeking.

Remark: The first part of the statement, that is, the equivariant embedding, was first discovered by Rosenlicht ([Ro], lemma 2 pp. 217).

Several constructions depending on a variety X (e.g. deformations) can be more easily described in terms of representation theory by the use of this proposition.

1.4. Every k-morphism $F : X \to W$ of a k-scheme X into a finite-dimensional k-vector space W can be interpreted in a natural way as an element of $k[X] \otimes W$ (i.e. we have $\mathrm{Mor}_k(X,W) \simeq \mathrm{Hom}_{k\text{-alg}}(S^*(W),k[X]) \simeq \mathrm{Hom}_{k\text{-vect}}(W,k[X]) \simeq k[X] \otimes W$; by the choice of a basis (w_i) in W one can write F in the form $\Sigma f_i \otimes w_i$, where the (f_i) are the coordinate functions of F corresponding to (w_i)). If a k-group acts regularly on X and linearly on W, then it operates linearly on $k[X] \otimes W$. The morphism F is G-equivariant exactly when it is a fixed point in $k[X] \otimes W$ under G. If M is a linear G-module, we designate the set of fixed points under G as M^G. Now assume that $X = V$, and W are finite-dimensional k-vector spaces on which G operates linearly. The differential $DF : V \to \mathrm{Hom}_k(V,W)$ of $F : V \to W$ is then an element of $k[V] \otimes V^* \otimes W$, and it follows from the chain rule that when F is equivariant, DF lies in $(k[V] \otimes V^* \otimes W)^G$:

$$DF = D(g \circ F \circ g^{-1}) = g \circ (D_{g^{-1}} F) \circ g^{-1} = g \cdot DF \ .$$

<u>1.5. Linearly Reductive Groups.</u> A finite-dimensional representation of the k-group G , i.e. a linear regular action on a k-vector space V , is called <u>completely reducible</u> when for every G-k-submodule V' of V , there is a complementary G-k-submodule V'' :

$$V = V' \oplus V'' \ .$$

A k-group G whose finite-dimensional representations are all completely reducible is called <u>linearly reductive</u>. A representation of G on a possibly infinite-dimensional k-vector space V is called <u>locally finite</u> when every element of V is contained in a finite-dimensional submodule. For example if G acts regularly on an affine k-scheme Spec R, the corresponding action of G on R is locally finite. If a representation of a reductive k-group G on a k-vector space V is locally finite, then V decomposes into a direct sum of irreducible G-k-modules. If one collects equivalent irreducible representations into isotypical components, then the decomposition of V into a direct sum of these isotypical components is unique. In particular, there is a distinguished complement $V^{\perp G}$ to the trivial G-k-submodule of fixed points V^G :

$$V = V^G \oplus V^{\perp G}$$

Nagata gives the following description of linearly reductive groups ([Na]):

In char(k) = 0 , all groups which have reductive identity components in the sense of [Bo] 11.21 , in particular all semisimple and finite groups, are linearly reductive. On the other hand, for char(k) = $p \neq 0$, only such groups G are linearly reductive for which the identity component G° is a torus and the quotient group G/G° has order prime to p .

2. Deformation Theory

2.1. Deformations. ([Ar4], [K-S], [Rim1], [Schl], [Tj1])

The notion of a family of varieties with sufficiently reasonable continuity properties is made precise in the following way.

Definition: A family of varieties is a flat k-morphism $\xi : X \to B$ of k-varieties. The variety X is called the total space and B the base of the family ξ .

Example: Consider $\xi : \mathbb{A}^2 \to \mathbb{A}^1$ given by $\xi(x,y) = xy$. This is a flat morphism whose fibers are plane curves.

$\xi^{-1}(t)$, $t \neq 0$

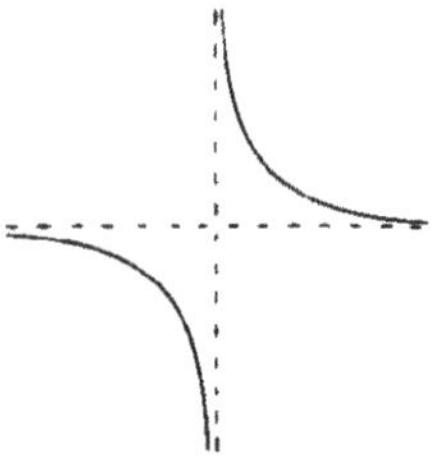

$\xi^{-1}(o)$

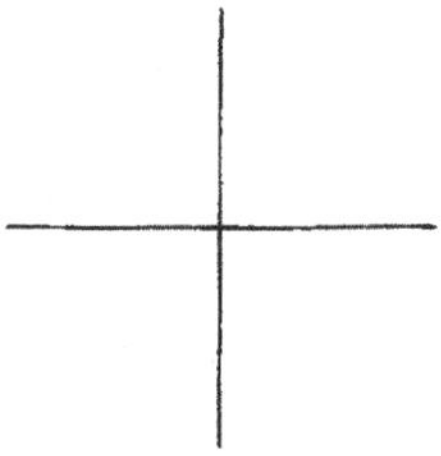

To put emphasis on a special fiber, like the singular one in this example, one defines the notion of deformation.

By a pointed k-variety (B,b) we will understand a k-variety B together with a k-rational point $b \in B$. A morphism $\phi : (C,c) \to (B,b)$ of pointed k-varieties is a k-morphism ϕ of the underlying varieties mapping c to b .

Definition: A deformation of a k-variety X_o consists of a family $\xi : X \to (B,b)$ over a pointed base and a k-isomorphism $j : X_o \to \xi^{-1}(b)$ of X_o onto the fiber over b . In the following we will denote the isomorphism j simply by an inclusion $X_o \hookrightarrow X$.

A <u>morphism</u> $(\phi,\Phi) : \eta \to \xi$ of two deformations $\xi : X \to (B,b)$, and $\eta : Y \to (C,c)$ of a variety X_o consists of a morphism $\phi : (C,c) \to (B,b)$ of pointed varieties, and a k-morphism $\Phi : Y \to X$ such that the following diagram is cartesian

$$\begin{array}{ccc} Y & \xrightarrow{\ \Phi\ } & X \\ {\scriptstyle\eta}\downarrow & & \downarrow{\scriptstyle\xi} \\ (C,c) & \xrightarrow{\ \phi\ } & (B,b) \end{array}$$

and

$$\begin{array}{ccc} & X_o & \\ \swarrow & & \searrow \\ Y & \xrightarrow{\ \Phi\ } & X \end{array}$$

commutes.

The condition that the first diagram be cartesian means that the natural morphism from Y to the fiber product $X \times_B C$ is an isomorphism. Thus a fiber $\eta^{-1}(d)$, $d \in C$, of the family η is isomorphic to the fiber $\xi^{-1}(\phi(d))$ of the family ξ . We also say that η is <u>induced</u> from ξ . Later on we wish to construct a universal deformation of a variety from which any other deformation of this variety may be induced. This requires a slight modification of the concepts introduced above.

<u>2.2. Formal Deformations</u> (cf. loc. cit., especially [Rim1]).

Let $\xi : X \to (B,b)$ be a deformation of a variety X_o . We are only interested in "small" deformations of X_o , and therefore it is reasonable to localize ξ with respect to b in B . Technically we have to go even further and to replace ξ by its prolongation $\hat{\xi} : \hat{X} \to \hat{B}$ on the formal completions of X and B along X_o and b (cf. EGA I 10.9). More generally we define:

<u>Definition:</u> A <u>formal deformation</u> of a k-variety X_o consists of a flat morphism

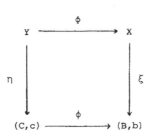

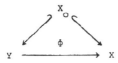

$\xi : X \to B$ from a formal k-scheme X to the formal spectrum $B = \mathrm{Spf}(R)$ of a complete local Noetherian k-algebra R with residue field k, and an isomorphism $j : X_o \to X \times_B \mathrm{Spec}\, k$ of X_o onto the fiber of ξ over the closed point of B.

For consistency with the definitions of 2.1 we regard B as canonically pointed by its closed point b corresponding to the residue homomorphism $B \to k$.

Let $\underline{m} \subset R$ be the maximal ideal of R, and for $q \in \mathbb{N}$ let $B_q := \mathrm{Spec}\, R/\underline{m}^{q+1}$ denote the q - th infinitesimal neighborhood of b in B. Put $X_q := X \times_B B_q$ and let $\xi_q : X_q \to B_q$ be the natural morphism induced by ξ. From the fact that X_o is a variety and the lemma below it follows that ξ is a morphism of finite type in the sense of EGA I 10.13, and, especially, that the ξ_q are morphisms of usual schemes whose inductive limit coincides with ξ.

Lemma: Let R be either an arbitrary ring with a nilpotent ideal $\underline{m}$, or a Noetherian local ring complete with respect to the topology defined by its maximal ideal $\underline{m}$, and let $u : M \to N$ be an R-module homomorphism with N flat. Then u is an isomorphism exactly when $u/\underline{m} : M/\underline{m}M \to N/\underline{m}N$ is one.

For a proof cf. EGA O, 6.6.21/22 or [Schl] Lemma 3.3.

A morphism of formal deformations $(\phi,\Phi): \xi \to \eta$ is similarly defined as in 2.1, usual morphisms being replaced by morphisms of formal schemes. The requirement that the square of the definition in 2.1 be cartesian may be replaced here by simply demanding commutativity. The lemma above then implies that the square is automatically cartesian. Especially, (ϕ,Φ) is an isomorphism exactly when ϕ is.

If $k = \mathbb{C}$ one may regard a variety as a complex analytic space. A reasonable notion of deformation is then obtained by replacing varieties by their germs at points, and morphisms by complex analytic morphisms (cf. [K-S], [Tj1]).

2.3. Semiuniversal Deformations. (Cf. loc. cit.)

Let X_o be a k-variety.

Definition: A formal deformation $\xi : X \to B$ of X_o is versal if for any other formal deformation $\eta : Y \to C$ of X_o there exists a morphism $(\phi,\Phi) : \eta \to \xi$. It is called semi-universal if in addition the differential $D_c\phi : T_cC \to T_bB$ from the tangent space of C in its closed point c to the tangent space of B in its closed point b is uniquely determined. (By "tangent space" we mean "Zariski tangent space"). If $\xi : X \to (B,b)$ is a deformation of X_o in the sense of 2.1 we say that it is versal resp. semi-universal if the corresponding completion $\hat{\xi}$ is. If it exists, a formal semi-universal deformation is uniquely determined up to isomorphism. This follows easily from the uniqueness condition on the differential $D_c\phi$ (cf. [Schl] 2.9).

Versality is most practically established inductively. In the following we use the notation $\eta_q : Y_q \to C_q$ for the restriction of a formal deformation $\eta : Y \to C$ to its q - th infinitesimal neighborhood (cf. 2.2).

Definition: A formal deformation ξ of X_o is infinitesimally versal if for any other formal deformation η of X_o, any $q \in \mathbb{N}$, and any morphism $\alpha_q : \eta_q \to \xi$ there is a morphism $\alpha_{q+1} : \eta_{q+1} \to \xi$ lifting α_q, i.e. fulfilling $\alpha_{q+1} \circ i = \alpha_q$ where i denotes the natural embedding of η_q into η_{q+1}.

Proposition: If a formal deformation is infinitesimally versal then it is versal.

Proof: Let $\xi : X \to B$ be an infinitesimally versal and $\eta : Y \to C$ an arbitrary formal deformation of a variety X_o. We obtain a morphism $(\phi,\Phi) : \eta \to \xi$ by taking the inductive limit of a sequence of lifting morphisms $(\phi_q,\Phi_q) : \eta_q \to \xi$, $q \in \mathbb{N}$, whose existence follows from the infinitesimal versality of ξ and the trivial start $(\phi_o,\Phi_o) : \eta_o \to \xi$ given by the inclusion of $\eta_o = \xi_o : Y_o = X_o \to \operatorname{Spec} k$ into ξ.

2.4. The Existence of Semiuniversal Deformations. In case a variety X_o has only isolated singularities a formal semiuniversal deformation of X_o exists (cf. [Rim1] 4.5, [Schl]). In our later applications we have only to deal with X_o a hypersurface. More generally let X_o be a complete intersection, that is, the variety X_o is isomorphic to the fiber $f^{-1}(0)$ of a flat k-morphism $f : \mathbb{A}^n_k \to \mathbb{A}^p_k$. Assume that $X_o \times_{\mathrm{Spec}\,k} \mathrm{Spec}\,\bar{k}$ has only isolated singularities, where $\bar{k}$ denotes an algebraic closure of k . A semiuniversal deformation of X_o may then be constructed in the following way. Let $x_1,\dots,x_n$ resp. $y_1,\dots,y_p$ be the coordinates of $\mathbb{A}^n_k$ resp. $\mathbb{A}^p_k$, and $f_1,\dots,f_p$ be the components $y_i \circ f$, $i=1,\dots,p$ of f . Since the singularities of X_o are isolated, the images of the vectors $(\frac{\partial f_1}{\partial x_j},\dots,\frac{\partial f_p}{\partial x_j})$, $j=1,\dots,n$ generate a $k[X_o]$-submodule J in $k[X_o]^p$, which is a k-vector subspace of finite codimension. Let $b_1,\dots,b_r$ be the representatives of a k-basis for $k[X_o]^p/J$ in $k[x_1,\dots,x_n]^p = k[\mathbb{A}^n_k]^p$ and let $F : \mathbb{A}^{n+r}_k \to \mathbb{A}^p_k$ be the k-morphism defined by

$$(x_1,\dots,x_n,u_1,\dots,u_r) \to f(x) + \sum_{i=1}^{n} u_i b_i(x) \ .$$

Let $X = F^{-1}(0)$, and let $\xi : X \to \mathbb{A}^r_k$ be the composition of the embedding $X \to \mathbb{A}^{n+r}_k$ and the second projection $\mathbb{A}^{n+r}_k \to \mathbb{A}^r_k$.

Theorem: The k-morphism $\xi : X \to (\mathbb{A}^r_k, 0)$ is a semiuniversal deformation of $X_o \cong \xi^{-1}(0)$.

For a proof see [Ar4] or [Rim1] 4.14.

Remarks: 1) The morphism ξ is a deformation even in the stronger sense of 2.1. For general X_o one only obtains a morphism $\xi : X \to B$ of formal schemes, and it is difficult to exhibit an X which is a scheme over a henselian local scheme B (cf. [Ar4] discussing the algebraization theorem of Elkik).

2) A complex analytic analogue of the above theorem is proved in [K-S] and [Tj1]. Here X_o is a germ at x of a complex space with isolated singularity x , and ξ is a morphism of germs $(X,x) \to (\mathbb{C}^r, o)$. This result is stronger than the theorem above as it says that a morphism (ϕ,Φ) from another complex-analytic deformation $\eta : (Y,x) \to (C,c)$ of (X_o,x) to ξ may be chosen to be holomorphic. One obtains this sharpening by choosing the liftings of the morphisms $(\phi_q,\Phi_q) : \eta_q \to \xi$ carefully so that their limit is realized by a convergent power series.

Example: The example family in 2.1 is a semiuniversal deformation of its singular fiber.

2.5 Equivariant Deformations. Let G be an algebraic k-group acting regularly on a variety X_o. If in the definitions of 2.1, 2.2, 2.3 we replace the category of k-varieties (resp. formal k-schemes) and k-morphisms by that of k-varieties (resp. formal k-schemes) with regular G-action and G-equivariant k-morphisms we obtain analogous notions of G-deformations, i.e. those of formal, G-versal, G-semi-universal, and infinitesimally G-versal G-deformations. For a pointed G-variety (B,b) the point b will be a fixed point of G on B. Proposition 2.3 obviously holds for G-deformations too. We wish to prove an analogon to theorem 2.4.

Definition: A G-k-variety X_o is called a G-complete intersection if it is G-k-isomorphic to the fiber $f^{-1}(0)$ of a flat G-equivariant k-morphism $f : V \to W$ between finite-dimensional k-vector spaces V and W on which G acts linearly.

Theorem: Let X_o be a G-complete intersection such that $X_o \times_{\mathrm{Spec}\, k} \mathrm{Spec}\, \bar{k}$ has only isolated singularities, and let G be linearly reductive. Then a G-semiuniversal G-deformation $\xi : X \to (U,o)$ of X_o exists. Moreover ξ is a G'-semiuniversal G'-deformation for any linearly reductive subgroup $G' \subset G$.

Proof: We first consider a semi-universal deformation $\xi : X \to (U,o)$ of X_o as given by theorem 2.4 and show that it admits natural G-actions on its base (U,o) and its total space X with respect to which ξ is equivariant. Subsequently we show that ξ is infinitesimally G-versal. This will imply the G'-semiuniversality of ξ for all linearly reductive subgroups $G' \subset G$.

i) The actions of G on X and U.
Let X_o be given as the fiber $f^{-1}(0)$ of a flat G-equivariant k-morphism $f : V \to W$ of k-vector spaces $V \cong \mathbb{A}^n$ and $W \cong \mathbb{A}^p$ on which G acts linearly. We will repeat the construction of 2.4 from a new point of view. The ideal J of 2.4 may be regarded as a $k[X_o]$-submodule of $k[X_o] \otimes W$. It is also a G-submodule of $k[X_o] \otimes W$ as it

coincides with the image of the G-module homomorphism

$$Tf \; : \; k[V] \otimes V \to k[X_o] \otimes W$$

defined on elements $h \otimes v$, $h \in k[V]$, $v \in V$, by

$$Tf(h \otimes v) \; = \; i^*(h \cdot Df(v))$$

where the differential Df of f is regarded as a G-equivariant linear map $V \to k[V] \otimes W$, and where i^* is the restriction

$$i^* : k[V] \otimes W \to k[X_o] \otimes W$$
$$h \otimes w \mapsto h|_{X_o} \otimes w \; .$$

As G is linearly reductive and its action on $k[V] \otimes W$ is locally finite, the G-equivariant projection

$$k[V] \otimes W \xrightarrow{p} k[X_o] \otimes W \longrightarrow (k[X_o] \otimes W)/J = U$$

of $k[V] \otimes W$ onto the finitedimensional cokernel U of Tf admits a linear G-section $s : U \to k[V] \otimes W$. If we interpret s as a G-invariant element of $U^* \otimes k[V] \otimes W \subset k[V \times U] \otimes W$ we obtain by

$$F \; = \; f \; + \; s$$

a G-invariant element of $k[V] \otimes W + U^* \otimes k[V] \otimes W \subset k[V \times U] \otimes W$ corresponding to a G-equivariant morphism $F : V \times U \to W$.

Let X be the G-stable subvariety $F^{-1}(O)$ of $V \times U$, and let $\xi : X \to U$ be the G-equivariant composition of the embedding $X \to V \times U$ and the second projection $V \times U \to U$. According to 2.4 we obtain thus a semiuniversal deformation of X_o which in addition is equivariant with respect to the G-actions on X and U. (After the choice of a basis $e_1,\dots,e_r$ of U the tensor s is written $\sum_{i=1}^{r} u_i \otimes b_i$, where $u_1,\dots,u_r \in U^*$ is a dual basis and $b_i = s(e_i) \in k[V] \otimes W$.)

ii) Infinitesimal G-versality.

Let $\eta : Y \to C$ be an arbitrary formal G-deformation of X_o. We have to show the following lifting property (notations C_q, Y_q as in 2.2, 2.3): For any $q \in \mathbb{N}$ and any G-equivariant morphism $(\psi,\Psi) : \eta_q \to \xi$ there is a G-equivariant lifting $(\phi,\Phi) : \eta_{q+1} \to \xi$.

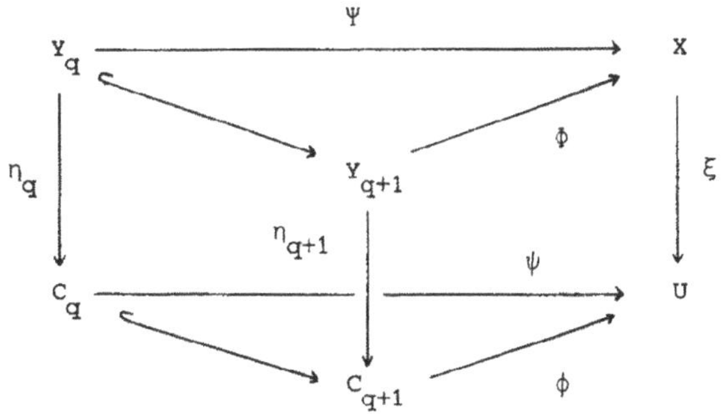

We regard Ψ as a morphism from Y_q to $V \times U$, that is as an element of $k[Y_q] \otimes (V \oplus U)$, whose U-component is $\psi \in k[C_q] \otimes U \subset k[Y_q] \otimes U$ (note that η_q is flat, hence $\eta_q^* : k[C_q] \to k[Y_q]$ is injective). We similarly do with Φ and ϕ.

Because of the ordinary semi-universality of ξ we may find a lifting Φ of Ψ, which possibly is not G-equivariant. As G is linearly reductive the G-module $k[Y_{q+1}] \otimes (V \oplus U)$ decomposes into the direct sum of its isotypical components. We may thus split $\Phi = \Phi^G + \delta\Phi$ into its equivariant part $\Phi^G \in (k[Y_{q+1}] \otimes (V \oplus U))^G$ and a pertubation $\delta\Phi$ lying in the G-complement $(k[Y_{q+1}] \otimes (V \oplus U))^{\perp G}$. Since Φ is a lifting of $\Psi = \Psi^G$ we have $\delta\Phi \in \underline{m}^{q+1} \cdot k[Y_{q+1}] \otimes (V \oplus U)$, where $\underline{m}$ denotes the maximal ideal of $k[C_{q+1}] \subset k[Y_{q+1}]$. Now Φ is a morphism from Y_{q+1} to $X \subset V \oplus U$ and must therefore fulfill the equation

$$F \circ \Phi = 0$$

(this equation being considered as an element of $k[Y_{q+1}] \otimes W$). The Taylor expansion of F gives

$$0 = F \circ \Phi = F \circ \Phi^G + (D_{\Phi^G}F)(\delta\Phi) .$$

Higher terms of this expansion do not occur, as they involve $\delta\Phi$ at least quadratically, and hence lie in $\underline{m}^{2q+2} \cdot k[Y_{q+1}] \otimes W = (0)$. The first term $F \circ \Phi^G$ lies in $(k[Y_{q+1}] \otimes W)^G$, and the second one is the image of $\delta\Phi$ under the G-module homomorphism

$$t : k[Y_{q+1}] \otimes (V \oplus U) \longrightarrow k[Y_{q+1}] \otimes W$$

$$h \otimes z \longmapsto h \cdot (\Phi^G)^* (DF(z)) .$$

Here we consider the differential DF of F as a G-equivariant linear map $V \oplus U \to k[V \times U] \otimes W$, and $(\Phi^G)^*$ denotes the G-equivariant pull back

$$k[V \times U] \otimes W \longrightarrow k[Y_{q+1}] \otimes W$$

$$g \otimes w \longmapsto (g \circ \Phi^G) \otimes w .$$

As G is linearly reductive, t respects the isotypical decomposition of the modules in question. Therefore $(D_{\Phi^G}F)(\delta F)$ lies in $(k[Y_{q+1}] \otimes W)^{\perp G}$, and for the fulfillment of the equation $F \circ \Phi = 0$ this term has to vanish as well as $F \circ \Phi^G$. Now, because of $F \circ \Phi^G = 0$, we may take $\Phi^G : Y_{q+1} \to X$ as the desired G-equivariant lifting of $\Psi : Y_q \to X$.

Remarks: 1) As was remarked in 2.4, in the event that $k = \mathbb{C}$ the morphisms (ϕ_q, Φ_q) generated inductively from (ϕ_o, Φ_o) can be chosen so that their inductive limit (ϕ, Φ) is locally convergent. In passing from the chosen (ϕ_q, Φ_q) to their equivariant parts (ϕ_q^G, Φ_q^G) it only may be necessary to remove several terms of the Taylor expansion of ϕ_q in $k[C_q] \otimes U$ resp. Φ_q in $k[Y_q] \otimes V$. So the formal morphism (ϕ,Φ) can be chosen both locally convergent and equivariant (cf. [K-S], lemma 2; [Tj1]).

2) In case $G = G_m$ (the multiplicative group) the theorem was proven by H. Pinkham (see [Pi1], [Pi2]) without assuming X_o to be a G-complete intersection. Whereas part ii) of the proof above does not depend on the fact that X_o is a G-complete intersection, the first one does, and it is not possible to generalize Pinkham's method of the "equivariant" lifting of the relations of f to others than k-diagonalizable groups in case X_o is not a G-complete intersection. Part (i) for general X_o was meanwhile obtained by D. S. Rim ([Rim2]). His proof runs along the abstract lines of [Rim1].

3) An analogue to theorem 2.5 can be proven for unfoldings of differentiable functions in the sense of Arnol'd, Mather and Thom (see [Sl1]).

2.6 Deformations with Fixed Symmetries. Let H be a k-group acting regularly and faithfully on a k-variety X_o. We may regard H as a subgroup of the automorphism group of X_o, and we will denote the couple consisting of X_o and the H-action on it simply by (X_o,H). By an automorphism of (X_o,H) we mean an automorphism of the k-variety X_o commuting with the action of H.

Let G be a k-group of automorphisms of (X_o,H). We may thus regard X_o as a $G \times H$-variety.

Definition: A (formal) G-deformation of (X_o,H) is a (formal) $G \times H$-deformation of X_o with trivial action of H on its base. A formal G-deformation $\xi : X \to B$ of (X_o,H) is called G-semiuniversal when for any other formal G-deformation $\eta : Y \to C$ of (X_o,H) there is a $G \times H$-equivariant morphism $(\phi,\Phi) : \eta \to \xi$ with uniquely determined differential $D_c\phi : T_cC \to T_bB$.

Now let $G \times H$ be linearly reductive and X_o a $G \times H$-complete intersection with isolated singularities. Then a $G \times H$-semi-universal $G \times H$-deformation $\xi : X \to U$ of

X_o exists (2.5). Denote by ξ^H the restriction $X \times_U U^H \to U^H$ of ξ over the H-fixed subspace U^H of U. Then ξ^H is a G-deformation of (X_o, H) and all $G \times H$-equivariant morphisms $(\phi, \Phi) : \eta \to \xi$ from formal G-deformations $\eta : Y \to C$ to ξ factorize over the embedding $\xi^H \to \xi$ with unique differential $D_c\phi$. We therefore obtain from 2.5:

Corollary: *Let $G \times H$ be linearly reductive and X_o a $G \times H$-complete intersection with isolated singularities. Let $\xi : X \to U$ be a $G \times H$-semiuniversal $G \times H$-deformation of X_o. Then $\xi^H : X \times_U U^H \to U^H$ is a G'-semiuniversal G'-deformation of (X_o, H) for any linearly reductive subgroup $G' \subset G$.*

Remark: For $G = 1$ the notion of semiuniversal deformation of (X_o, H) is analogous to that of H-versal unfolding for a differentiable function as introduced by Poênaru (cf. [Po]).

2.7. Deformations with G_m-action. Let $G = G_m$ be the multiplicative group. We say G_m operates on the vector space $V \cong \mathbb{A}^n$ with weights $w_1, \dots, w_n \in \mathbb{Z}$ if for all $x = (x_1, \dots, x_n) \in V$ and $t \in G_m$ we have

$$t \cdot (x_1, \dots, x_n) = (t^{w_1} x_1, \dots, t^{w_n} x_n) .$$

Every linear action of G_m is of this form (cf. [Bo] § 8). If G_m operates on another vector space $W \cong \mathbb{A}^p$ with weights $d_1, \dots, d_p$, we call a G_m-equivariant map $f : V \to W$ quasi-homogeneous of type $(d_1, \dots, d_p ; w_1, \dots, w_n)$, cf. also 7.4.

Now let $f : V \to W$ be a flat G_m-equivariant morphism, such that $X_o = f^{-1}(0)$ is a G_m-complete intersection. Assume that X_o has only isolated singularities. The cokernel U of Tf (in the proof of 2.5 i)) decomposes under the G_m-action into a direct sum of onedimensional irreducible G_m-modules $U = \bigoplus_{i=1}^{r} U_i$, on which G_m acts respectively with certain weights $m_1, \dots, m_r \in \mathbb{Z}$.

Let M be a not necessarily finite-dimensional G_m-module, which decomposes into a direct sum of finite-dimensional G_m-eigenspaces $M(j)$ for the weights $j \in \mathbb{Z}$. The formal expression

$$P_M(T) = \sum_{j \in \mathbb{Z}} \dim_k M(j) T^j$$

is called the <u>characteristic function</u> of the G_m-module M.

In the above situation, let $p = 1$, i.e. $\dim W = 1$, and let the weights for G_m on V and W be positive. Then $f : V \to W$ is given by a quasihomogeneous polynomial of type $(d;w_1,\dots,w_n)$ and the Euler identity

$$d \cdot f = \sum_{i=1}^{n} w_i x_i \frac{\partial f}{\partial x_i}$$

holds. If the characteristic of k doesn't divide the degree d of f, then the ideal I in $k[V]$ generated by f and the partial derivatives $\frac{\partial f}{\partial x_i}$, $i=1,\dots,n$, can be generated by the $\frac{\partial f}{\partial x_i}$, $i=1,\dots,n$ alone. As these partial derivatives $\frac{\partial f}{\partial x_i}$ are quasihomogeneous of type $(d-w_i;w_1,\dots,w_n)$, the subalgebra A generated by them is a G_m-submodule of $k[V]$. For any G_m-complement U' of I in $k[V]$ we have, because the singularities of $f^{-1}(0)$ are isolated, that $k[V] = U' \otimes_k A$, and the characteristic function $P_U(T) = P_{U'}(T)$ of the base U of the G_m-semi-universal deformation of $f^{-1}(0)$ satisfies the relation

$$T^d P_U(T^{-1}) = \prod_{i=1}^{n} \frac{(1-T^{d-w_i})}{(1-T^{w_i})} .$$

(for details cf. [A3] § 4.5 or LIE V § 5, n° 5 Lemme 5, n° 1 Prop 2). Therefore the G_m-action on U is completely determined by the quasihomogeneous type $(d;w_1,\dots,w_n)$. In particular we have $\dim_k U = P_U(1)$.

<u>2.8. Deformations of Completions.</u> Up to now we have considered deformations of varieties X_o. We may equally well substitute a formal scheme itself at the place of X_o in the definition of formal deformation (cf. 2.2). A notion of semi-universal deformation is then obtained by the analogous definition (2.3).

Let $\xi : X \to (B,b)$ be a deformation of a k-variety X_o, and let $x \in X_o$ be a k-rational point. Then the prolongation $\bar{\xi} : \bar{X} \to \bar{B}$ of ξ to the completions $\bar{X}$ and $\bar{B}$ of X in x and B in b will be a formal deformation of the completion $\bar{X}_o$ of X_o in x.

Theorem: Suppose that x is the only singular point of X_o. Then ξ is semiuniversal exactly when $\bar{\xi}$ is.

Proof: If ξ is semiuniversal then the semiuniversality of $\bar{\xi}$ follows from [Rim1] 4.10 (in case that X_o is a complete intersection one can also use the formal parts of the analytic proofs of 2.4 in [K-S] and [Tj1]). Conversely let $\bar{\xi}$ be semiuniversal, and let $\eta : Y \to C$ be a formal semiuniversal deformation of X_o. Then $\hat{\xi}$ (cf. 2.2) can be induced from η by a morphism $(\phi,\Phi) : \hat{\xi} \to \eta$. Its prolongation $(\phi,\bar{\Phi}) : \bar{\xi} \to \bar{\eta}$ to $\bar{\xi}$ and $\bar{\eta}$ must be an isomorphism as $\bar{\eta}$ is semiuniversal, according to the first part. But this means that ϕ is an isomorphism which implies that $(\phi,\Phi) : \hat{\xi} \to \eta$ is an isomorphism too (cf. Lemma 2.2). Hence $\hat{\xi}$, and then by definition ξ are semiuniversal deformations of X_o.

Corollary: Let $\xi : X \to (B,b)$ resp. $\eta : X \to (B,b)$ be deformations of k-varieties X_o resp. Y_o with single singular points $x \in X_o$ resp. $y \in Y_o$. Let the completions $\bar{X}$ and $\bar{Y}$ of X in x and of Y in y be isomorphic over B. Then ξ is semiuniversal exactly if η is.

Remark: In the case of complete intersections the corollary can also be proved by a direct calculation.

II Conjugacy Classes and Singularities in Semi-simple Groups

In this part, k will be an algebraically closed commutative field.

3. The quotient of the adjoint action

3.1. Reductive Groups ([Bo] § 14, [St2] 3.1 - 3.4).

Let G be an algebraic group over k. Then G will operate regularly on itself by means of the inner automorphisms $g \in G \mapsto (\text{Int } g : G \to G,\ x \mapsto {}^g x = gxg^{-1},\ x \in G)$. This action of G is called the adjoint action. The linearization of the adjoint action we will call the adjoint representation, i.e. the action of G on the Lie algebra $\underline{g}$ of G by means of the differentials $\text{Ad } g = D_e \text{ Int } g : \underline{g} \to \underline{g}$ of the automorphisms $\text{Int } g$. Now let G be reductive (and so by definition connected), and let $T \subset G$ be a maximal torus of G with character group $X^*(T) \cong \mathbb{Z}^r$, where r is the rank of G. Let $N(T)$ be the normalizer of T in G. Then the group $W = N(T)/T$ is finite and is called the Weyl group of G (with respect to T). If we restrict the adjoint representation of G to T, the Lie algebra $\underline{g}$ of G will decompose into a direct sum

$$\underline{g} = \bigoplus_{\alpha \in X^*(T)} \underline{g}_\alpha$$

of eigenspaces $\underline{g}_\alpha$, on which T acts by the character α. The elements of the finite set $\Sigma = \{\alpha \in X^*(T) \mid \alpha \neq 0,\ \underline{g}_\alpha \neq \{0\}\}$ are called the roots of T in $\underline{g}$ (and also G). For each $\alpha \in \Sigma$ the eigenspace $\underline{g}_\alpha$ is one-dimensional, and $\underline{g}_0$ is the Lie algebra of T. For every root $\alpha \in \Sigma$ there is an isomorphism $\phi_\alpha : G_a \to U_\alpha$ of the additive group G_a to a subgroup $U_\alpha \subset G$ normalized by T, such that the Lie algebra of U_α is exactly $\underline{g}_\alpha$ and

$$t\,\phi_\alpha(x)t^{-1} = \phi_\alpha(\alpha(t)\cdot x)$$

holds for all $t \in T$ and $x \in G_a$.

The reductive group G decomposes into the almost direct product $G = T'.G'$ of a central torus T' and the semisimple commutator group $G' = (G,G)$ of G. Now assume that G is semisimple. Then Σ determines a root system of rank r in the vectorspace $X^*(T)_{\mathbb{Q}} = X^*(T) \otimes_{\mathbb{Z}} \mathbb{Q}$. The Weyl group of this root system can be identified with W by the natural action of W on $X^*(T)_{\mathbb{Q}}$. Let $(,)$ be a W-invariant scalar product on $X^*(T)_{\mathbb{Q}}$, and let $L(\Sigma)$ be the $\mathbb{Z}$-lattice generated by the roots. The lattice of weights will be $L^*(\Sigma) = \{\omega \in X^*(T)_{\mathbb{Q}} | \frac{2(\omega,\alpha)}{(\alpha,\alpha)} \in \mathbb{N}$, for all $\alpha \in L(\Sigma)\}$, and we will have the inclusions

$$L(\Sigma) \subset X^*(T) \subset L^*(\Sigma) .$$

The quotient $\pi_1(G) = L^*(\Sigma)/X^*(T)$ is the fundamental group of G, and G is called simply connected resp. adjoint exactly when $X^*(T) = L^*(\Sigma)$ resp. $X^*(T) = L(\Sigma)$. The choice of a Borel subgroup B of G containing T is equivalent to the choice of a system of positive roots Σ^+ or of a basis Δ of the root system Σ. The image of the product map $T \times \prod_{\alpha \in \Sigma^+} U_\alpha \to B$ is then an isomorphism of varieties.

Any $\lambda \in X^*(T)$ which satisfies $\frac{2(\lambda,\alpha)}{(\alpha,\alpha)} \in \mathbb{N}$ for all $\alpha \in \Delta$ is called a dominant weight. For every dominant weight λ there is an irreducible representation $\rho_\lambda : G \to GL(V_\lambda)$, unique up to isomorphism, on a finite-dimensional vector space V_λ, which has λ as its highest weight. We define $\chi_\lambda : G \to k$, $\chi_\lambda(g) = \text{Trace } \rho_\lambda(g)$ to be the character of ρ_λ. Then χ_λ is regular, i.e. an element of $k[G]$, and is constant on the conjugacy classes of G. If G is simply connected and $\Delta = \{\alpha_1,\dots,\alpha_r\}$, then because $L^*(\Sigma) = X^*(T)$, there exist fundamental dominant weights λ_i, $i=1,\dots,r$, with $\frac{2(\lambda_i,\alpha_j)}{(\alpha_j,\alpha_j)} = \delta_{ij}$. We denote by χ_i, $i=1,\dots,r$, the characters of the associated representations.

If G is simple, i.e. G contains no positive dimensional normal subgroup, then G is determined by the now irreducible root system Σ in $X^*(T)$ up to a central isogeny, i.e. a homomorphism with a finite central kernel. After the choice of a basis Δ of Σ the information given by Σ is equivalent to that given by the Cartanmatrix $((n_{\alpha,\beta}))$, $n_{\alpha,\beta} = \frac{2(\alpha,\beta)}{(\alpha,\alpha)}$, $\alpha,\beta \in \Delta$, or by the Dynkin diagram of Σ

(cf. LIE VI). The Dynkin diagrams corresponding to irreducible root systems can be classified by the following list (LIE VI, 4.2, Th. 3):

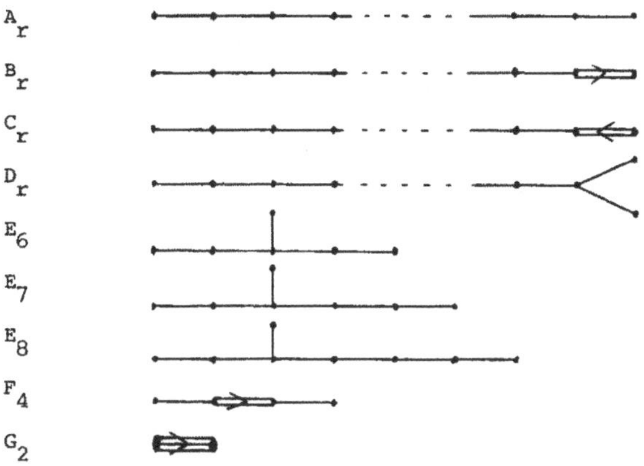

A_r ($r \geq 1$ vertices)

B_r ($r \geq 2$ vertices)

C_r ($r \geq 3$ vertices)

D_r ($r \geq 4$ vertices)

The vertices of the diagrams correspond to the simple roots of the basis $\Delta \subset \Sigma$, and the edges satisfy:

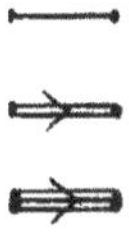

$n_{\alpha,\beta} = -1$

$n_{\alpha,\beta} = -2 \qquad \alpha \neq \beta \in \Delta$

$n_{\alpha,\beta} = -3$

In section 3.11 the previous and the following notions will be illustrated by means of the example $G = SL_n$.

3.2. Invariants on G. Let G be semisimple, T a maximal torus of G. The invariant functions $k[G]^G$ on G with respect to the adjoint action form a finitely generated subalgebra of $k[G]$ (cf. [M2] Th. 1.1, [Sp3] 2.4). Then we have:

Theorem ([St2] 3.4 Th. 2): The homomorphism $k[G]^G \to k[T]^W$ induced by the inclusion $T \subset G$ is an isomorphism, and the characters χ_λ, for dominant λ, resp. their restrictions $\chi_{\lambda|T}$ to T, are a k-basis of the vector space $k[G]^G$ resp. $k[T]^W$. If in addition G is simply connected, then $k[G]^G$, resp. $k[T]^W$, will be generated freely as a k-algebra by the χ_i, $i=1,\dots,r$, resp. their restrictions $\chi_{i|T}$.

The morphism $T \to \mathrm{Spec}\, k[T]^W$ induced by the inclusion $k[T]^W \to k[T]$ is surjective ([St2] 3.4, Cor. 2, or in general [M2] Chap. 0, § 2 (3), Th. 1.1). Therefore we can interpret the elements of $\mathrm{Spec}\, k[T]^W(k)$ as W-conjugacy classes $\bar{t} \in T/W$ of elements $t \in T$. The morphism $\chi : G \to \mathrm{Spec}\, k[G]^G$ induced by the inclusion $k[G]^G \to k[G]$ (also surjective, loc. cit.) is called the adjoint quotient of G, and it will be written in the form $\chi : G \to T/W$, the choice of a maximal torus T tacitly assumed.

The fibers of χ are unions of conjugacy classes of elements of G. An important aid for their investigation is the Jordan decomposition of elements of G.

3.3. Jordan decomposition ([Bo] I. 4.1-4, [St2] 2.1-4).
Let V be a finite-dimensional k-vector space, and let x be a k-linear endomorphism of V. Then call x semisimple resp. nilpotent resp. unipotent, when x is diagonalizable resp. when there exists $m \in \mathbb{N}$ with $x^m = 0$ resp. when there exists $m \in \mathbb{N}$ with $(x\text{-}id)^m = 0$. If G is an (affine) algebraic group and $\rho : G \to GL(V)$ is a faithful representation of G (cf. 1.1 and [St2] 1.9), then call an element $x \in G$ semisimple resp. unipotent when $\rho(x)$ is semisimple resp. unipotent. This definition doesn't depend on the choice of a representation ρ (there exists an invariant definition, cf. loc. cit.), and if $\phi : G \to G'$ is a morphism of algebraic groups, $\phi(x)$ will be semisimple resp. unipotent when x is semisimple resp. unipotent. Every element $x \in G$ can be decomposed in a unique way $x = x_s \cdot x_u$, where x_s is semisimple, x_u is unipotent, and x_s, x_u commute. Morphisms of algebraic groups respect the Jordan decomposition by its uniqueness. To determine the decomposition of an element $x \in G$ it suffices to know the situation in the general linear groups GL_n, $n \in \mathbb{N}$. Let $x \in GL_n$, then there are polynomials $P, Q \in k[T]$ in one indeterminate with vanishing constant which represent x_s, x_u, i.e. such that $x_s = P(x)$, $x_u = Q(x)$. These polynomials depend only on the characteristic polynomial of x.

Remark: Steinberg uses (loc. cit.) the minimal polynomial of x. However for us it will be important to use the characteristic polynomial as in Borel ([Bo] I. 4.2). In general the function $x \mapsto x_s$ is not a morphism. Consider the example SL_2. Let $x \in SL_2$ be the matrix $\begin{pmatrix} a & b \\ c & d \end{pmatrix}$, so that the characteristic polynomial of x has the form $T^2-(a+d)T+1$. Then x is unipotent resp. semisimple when $a+d = 2$ resp. $\neq \pm 2$. The function $x \mapsto x_s$ is the identity on the open set $a+d \neq \pm 2$ of SL_2. If it were a morphism, it would then be the identity on all of SL_2 in contradiction to the fact that unipotent elements exist in SL_2 (for example $\begin{pmatrix} 1 & 1 \\ 0 & 1 \end{pmatrix}$).

In the investigation of the adjoint action of G, the structure of the isotropy group of a point $x \in G$, i.e. the centralizer $Z_G(x) = \{g \in G \mid gxg^{-1} = x\}$, is of interest. Since the semisimple part x_s and the unipotent part x_u are polynomials in x, it follows immediately that $Z_G(x) = Z_G(x_s) \cap Z_G(x_u)$. Or, as x_u lies in $Z_G(x_s)$ we obtain $Z_G(x) = Z_{Z_G(x_s)}(x_u)$.

Let $x = x_s \cdot x_u$ and $y = y_s \cdot y_u$ be the Jordan decomposition of two elements $x,y \in G$ which have $x_s = y_s$. It follows directly from the uniqueness of the decomposition that x and y are conjugate in G exactly when x_u and y_u are in $Z_G(x_s)$. Therefore we next consider the group $Z_G(x_s)$.

<u>3.4. Centralizers of semisimple elements.</u> Let the group G be reductive. If x is semisimple, x will lie in a maximal torus T of G ([St2] 2.13, Th. 1).

<u>Theorem</u> ([St2] 3.5, Prop. 4, [S-S] II.3.9, 4.1, 4.4):
<u>Let</u> $t \in T$, $\Sigma_t = \{\alpha \in \Sigma \mid \alpha(t) = 1\}$ <u>be the system of roots which vanish at</u> t, <u>and let</u> $W' = Z_W(t)$ <u>be the stabilizer of</u> t <u>in</u> W. <u>Then the centralizer</u> $Z_G(t)$ <u>will be generated by</u> T, <u>the root subgroups</u> U_α <u>with</u> $\alpha \in \Sigma_t$, <u>and a system of representatives of</u> W' <u>in</u> $N(T)$. <u>The identity component</u> $Z_G(t)^\circ$ <u>is reductive of the same rank as</u> G, <u>and</u> Σ_t <u>is the root system of its semisimple commutator. If</u> G <u>is semisimple and simply connected, then</u> $Z_G(t)$ <u>is connected.</u>

It can be shown that the unipotent part x_u of an element $x \in G$ lies in $Z_G(x_s)^\circ$ ([St2] 2.13, Cor. 4). Then x_u also lies in the semisimple commutator of $Z_G(x_s)^\circ$. With that, the study of the conjugacy classes of G reduces (see also the Remark at the end of 3.3) to the investigation of the unipotent conjugacy classes in "smaller" semisimple subgroups of G , about which we will make several statements.

3.5 Rationally closed root subsystems. (We keep the same notation as in 3.4.)

Definition: Let R be a subring of $\mathbb{Q}$. A subset $\Sigma' \subset \Sigma$ is called R-closed when all elements of Σ which are R-linear combinations of elements of Σ' already lie in Σ' .

Example: The systems consisting of the short roots of type $A_1 \times A_1$ in B_2 or of type A_2 in G_2 are not $\mathbb{Z}$-closed. The systems consisting of the long roots of type $A_1 \times A_1$ in B_2 and of type A_2 in G_2 are $\mathbb{Z}$-closed, however not $\mathbb{Q}$-closed. (Analogous statements hold in B_r , C_r , F_4 .)

Let G be semisimple and T a maximal torus of G . For all $t \in T$, the system $\Sigma_t = \{\alpha \in \Sigma | \alpha(t) = 1\}$ is obviously $\mathbb{Z}$-closed in Σ . With regard to future applications, we prove the following partial sharpening of the result:

Lemma: There exists a W-stable open neighborhood Q of the identity $e \in T$, such that for all $t \in Q$, $t \neq e$, Σ_t is either a proper $\mathbb{Q}$-closed root subsystem of Σ , or is empty.

Proof: Let $t \in T$ such that Σ_t is not $\mathbb{Q}$-closed. Then t lies in one of the connected components of the closed subgroup $S = \{s \in T | \alpha(s) = 1$ for all $\alpha \in \Sigma_t\}$ which doesn't contain the identity e , as all rational linear combinations of Σ_t that lie in $X^*(T)$ or Σ vanish on the identity component S° (cf. [St2] 2.6, [Bo] III 8.5, 8.7). The desired open set Q is obtained from T by removing all the connected components of the finitely many subgroups $T_{\Sigma'} = \{t \in T | \alpha(t) = 1$ for all $\alpha \in \Sigma'\}$, $\Sigma' \subset \Sigma$, which do not contain the identity. The subsystems $\Sigma_t \subset \Sigma$ for $t \in Q$, $t \neq e$ will be contained properly in Σ , since $\Sigma_t = \Sigma$ implies that t lies in the center of G , which is discrete. The W-stability of Q is obvious.

Proposition (LIE VI § 1, n° 1.7, Prop. 24):
If $\Sigma' \subset \Sigma$ is a $\mathbb{Q}$-closed root subsystem of Σ, then every basis of Σ' can be extended to a basis of Σ.

Corollary: The Dynkin diagram of a $\mathbb{Q}$-closed root subsystem $\Sigma' \subset \Sigma$ arises from the diagram of Σ by the removal of vertices and all associated edges.

Example: All proper $\mathbb{Q}$-closed root subsystems of A_3 are of type A_2, A_1 or $A_1 \times A_1$.

3.6. Torsion. (This section will be needed only for fields of positive characteristic. The earlier notation will be retained.) Let the group G be semisimple. For the sake of brevity, and without fear of a misunderstanding, we will improperly use $\pi_1(R)$ for the fundamental group $\pi_1((R,R))$ of the semisimple commutator of a reductive group R. Besides the type of the root system Σ_t of $Z_G(t)$, the order of $\pi_1(Z_G(t))$ will also be of interest later. That is because for every semisimple group H there is a central isogeny $\Phi : \tilde{H} \to H$ from the simply connected group $\tilde{H}$ having the same root type as H. This isogeny is separable (or étale) exactly when the characteristic of k does not divide the order of $\pi_1(H)$. In that case, many statements about H reduce to statements about $\tilde{H}$. We will carry out such reductions for the semismiple commutator of the group $Z_G(t)$ in 4.4.

Next, recall the calculation of the fundamental group $\pi_1(Z_G(t))$ (cf. [St3] 2.13). Consider the inversion $\alpha \mapsto \alpha^* = \dfrac{2\alpha}{(\alpha,\alpha)}$, $\alpha \in X^*(T)_{\mathbb{Q}}$, on the rational character group $X^*(T)_{\mathbb{Q}}$ of a maximal torus T of the group G. The image Σ^* of Σ under this map is a root system dual to Σ. Let $\Sigma_t \subset \Sigma$ be the root subsystem belonging to $Z_G(t)$, $t \in T$, and let $L(\Sigma^*)$ resp. $L(\Sigma_t^*)$ be the $\mathbb{Z}$-lattice in $X^*(T)_{\mathbb{Q}}$ generated by Σ^* resp. Σ_t^*. Then there is a homomorphism $\pi_1(Z_G(t)) \to \pi_1(G)$, whose kernel is isomorphic to the torsion of the quotient $L(\Sigma^*)/L(\Sigma_t^*)$.

Example: Let $Q \subset T$ be the open neighborhood of the identity found in Lemma 3.5. For all $t \in Q$, Σ_t is a $\mathbb{Q}$-closed root subsystem of Σ, and it follows from [St3]

2.11 that $L(\Sigma^*)/L(\Sigma_t^*)$ is torsion-free. Therefore the homomorphism $\pi_1(Z_G(t)) \to \pi_1(G)$ is injective, and if G is simply connected, so is the commutator of $Z_G(t)$.

A reductive subgroup G' of G is called very regular when it contains a maximal torus T of G, and the root system Σ' of T in G' is a $\mathbb{Z}$-closed root subsystem of the system Σ of T in G. By the remark in 3.5, the groups $Z_G(t)$ are very regular.

Definition: A prime p is called a torsion prime of G, when there exists a very regular subgroup G' of G, such that p divides the order of its fundamental group. A prime p is called a torsion prime of a root system Σ, when there exists a $\mathbb{Z}$-closed root subsystem Σ' of Σ, such that $L(\Sigma^*)/L(\Sigma'^*)$ has torsion of order p.

Lemma ([St3] 2.5): The prime p is a torsion prime for the group G exactly when it divides the order of $\pi_1(G)$ or it is a torsion prime of the root system Σ of G.

The fundamental group $\pi_1(G)$ is a quotient of the biggest possible fundamental group $L^*(\Sigma)/L(\Sigma)$ for the root system Σ of G. The following list contains the order of these groups and the torsion primes for irreducible systems Σ ([St3] 1.12, LIE VI § 4 n° 4):

Σ	A_r	B_r	C_r	D_r	E_6	E_7	E_8	F_4	G_2
Torsion Primes	/	2	/	2	2,3	2,3	2,3,5	2,3	2
$\# L^*(\Sigma)/L(\Sigma)$	r+1	2	2	4	3	2	1	1	1

3.7. Associated fiber bundles. Let G be an algebraic group and $H \subset G$ a closed subgroup. The quotient $G \to G/H$ is then a principal fiber bundle over the base G/H with structure group H in the sense of Serre ([Se] 2.2, 2.5). If F is a reduced variety (a restriction made for the sake of simplicity and to conform with [Se], however see [De-Ga] III § 4, n° 3 for a generalization), on which H operates regularly, we define for $G \to G/H$ the <u>associated fiber bundle</u> $G \times^H F$ <u>with fiber</u> F as the quotient (in the sense of [Bo] II.6.1) of $G \times F$ by the H-action $H \times G \times F \to G \times F$, $(h,g,f) \mapsto (gh^{-1},hf)$, (cf. [Se] 3.2). The regular functions on $G \times^H F$ can be identified with the H-invariant regular functions on $G \times F$, i.e. we have $k[G \times^H F] = k[G \times F]^H$.

The operation of G on itself by left translations induces in a natural way regular actions of G on G/H, $G \times F$ and $G \times^H F$ ([Bo] 6.10, 6.13, [Ro] Prop. 2).

We write $g * f$ for the class of $(g,f) \in G \times F$ in $G \times^H F$. The morphism $g * f \mapsto gH$ of the bundle $G \times^H F$ to the base G/H is equivariant with respect to G. (Locally, it is also easy to see that the morphisms $G \times F \to G \times^H F$, $G \times^H F \to G/H$ are separable.)

Let E be a second (reduced) variety with an H-action, and $\Phi : E \to F$ an H-equivariant morphism. The morphism $id \times \Phi : G \times E \to G \times F$ will induce a G-equivariant morphism $G \times^H \Phi : G \times^H E \to G \times^H F$. If F is a vector space with linear H-action resp. a group on which H acts by means of group automorphisms, then $G \times^H F$ is a vector bundle resp. a group bundle.

The following simple lemma will be useful later.

<u>Lemma 1</u>: <u>Let the</u> H-<u>action on the variety</u> F <u>be the restriction of a</u> G-<u>action on</u> F <u>and let</u> G <u>operate on</u> $G/H \times F$ <u>diagonally over both factors. Then the map</u> $\tau : G \times^H F \to G/H \times F$ <u>defined by</u> $\tau(g * f) = (gH, gf)$ <u>is a</u> G-<u>equivariant isomorphism</u>.

<u>Proof</u>: The morphism $G \times F \to G/H \times F$, $(g,f) \mapsto (gH,gf)$ factors through τ. Similarly, $G \times F \to G \times^H F$, $(g,f) \mapsto g * g^{-1}f$ factors through a morphism $G/H \times F \to G \times^H F$, which is an inverse to τ.

If $E \subset F$ is an H-stable (reduced) subvariety of F, where F is as in Lemma 1, the isomorphism τ above gives a G-equivariant embedding

$$G \times^H E \hookrightarrow G \times^H F \xrightarrow{\sim} G/H \times F$$

$$g * e \longmapsto (gH, ge) .$$

Lemma 2: Let $H \subset K$ be algebraic subgroups of G, K operating on F. Then $\kappa : G \times^H F \to G \times^K (K \times^H F)$ defined by $\kappa(g * f) = g * 1 * f$ is a G-isomorphism (where 1 is the identity in K).

Proof: The morphism $G \times F \to G \times^K (K \times^H F)$, $(g,f) \to g * 1 * f$, factors through κ, and $G \times K \times F \to G \times^H F$, $(g,k,f) \to gk * f$, factors through κ^{-1}.

Given the assumptions of Lemma 2, it follows directly from the definition and [Bo] III. 6.10, 6.13 that:

Lemma 3: If H is normal in G, and operates trivially on F, then $G \times^K F$ and $(G/H) \times^{(K/H)} F$ are G-isomorphic.

Lemma 4: Let $\Phi : X \to G/H$ be a G-equivariant morphism from the G-variety X to the homogeneous space G/H, and let $E \subset X$ be the fiber $\Phi^{-1}(H)$. Then E will be stabilized by H, and the map $\Psi : G \times^H E \to X$, $\Psi(g * e) = ge$, defines an isomorphism of G-varieties.

Proof: We will factor Ψ into an immersion and a projection using Lemma 1

$$G \times^H E \xrightarrow{\tau} G/H \times X \xrightarrow{\pi_2} X$$

$$g * e \longmapsto (gH, ge) \longmapsto ge$$

The map Ψ is obviously surjective (use the transitivity of G on G/H). Therefore it suffices to find a G-section $\gamma : X \to G/H \times X$ of π_2 onto the image of τ.

Since $\Phi(ge) = gH$, the graph of Φ , $\gamma(x) := (\Phi(x),x)$, is one such section.

As an example of an application of Lemma 4 which we will use later (in 6.4) we note:

Corollary: *Let* $U \triangleleft P \subset G$ *be subgroups of* G , P *acting on the normal subgroup* U *by conjugation. Let* $s : G/P \to G \times^P U$ *be a section of the associated bundle, and* N *be the normal bundle of* $s(G/P)$ *in* $G \times^P U$. *Then we have* $s^*(N) \cong G \times^P \mathrm{Lie}(U)$, *where* P *operates on* $\mathrm{Lie}(U)$ *by means of the adjoint representation.*

Proof: Fiberwise multiplication with a section s yields an isomorphism of $G \times^P U$ to itself as a variety which sends the identity section to s . So it suffices to consider s an identity section which goes into itself under the G-action. Consequently, G also operates on N , and the projection $N \to G/P$ is G-equivariant. By Lemma 4, it is enough to know the action of P in the fiber of N over $P \in G/P$. However that is just the differential at $e \in U$ of the conjugation by P on U , i.e. the adjoint representation on $\mathrm{Lie}(U)$.

Remark: Bundles in the sense of Serre ([Se]) are in general not locally trivial in the Zariski topology, rather only in the finer êtale topology (locally isotrivial). However there are criteria in the case of a reductive group G with subgroup H for the bundle $G \to G/H$ to be locally trivial in the Zariski topology (cf. [Bo-Ti] 3.24, 3.25). If, for example, $H = P$ a parabolic subgroup of G , then $G \to G/P$ is locally trivial.

3.8. Regular elements in reductive groups. Let G be a reductive group of rank r . An element $x \in G$ is called *regular* exactly when the centralizer $Z_G(x)$ of x in G has minimal dimension. The orbit of x under the adjoint action, i.e. the conjugacy class, will then have the maximal orbit dimension in G . In that case we also say that the orbit of x is regular. It is easy to see ([St2], 3.5 Prop. 1) that the minimal dimension of a centralizer is equal to the rank r of G .

The following characterizations of regular elements are due to Steinberg ([St1] 3.1,

3.3, 8.1, [St2] 2.6, Prop. 7, 3.5, Prop. 3, 3.6 Th. 3, 3.7 Th. 1, Th. 2, 3.8 Th. 3).

Theorem: An element x of a reductive group G is regular exactly when it is contained in only finitely many Borel subgroups of G. For x semisimple the following are equivalent:

$a)_s$ x is regular,

$b)_s$ x lies in exactly one maximal torus of G,

$c)_s$ if T is a maximal torus containing x, then $\alpha(x) \neq 1$ for all roots α of T in G.

In particular, there are regular semisimple elements in G. There is exactly one conjugacy class of regular unipotent elements in G, and for x unipotent in G, the following are equivalent:

$a)_u$ x is regular,

$b)_u$ x lies in exactly one Borel subgroup of G,

$c)_u$ if $B = T \cdot \prod_{\alpha \in \Sigma^+} U_\alpha$ is a Borel subgroup containing x and $x = \prod_{\alpha \in \Sigma^+} \phi_\alpha(c_\alpha)$ is a representation of x with $c_\alpha \in k$, then $c_\alpha \neq 0$ for all α in the basis Δ of Σ^+.

If G is semisimple and simply connected, then $x \in G$ is regular exactly when x is a regular point of the adjoint quotient $\chi : G \to T/W$, that is, when the differential $D_x \chi : T_x G \to T_{\chi(x)}(T/W)$ is surjective.

3.9. The unipotent variety. Let G be semisimple of rank r, and $x \in G$. If $\rho : G \to GL(V)$ is a representation of G, then by the choice of a suitable basis in V, $\rho(x)$ will represent an upper triangular matrix whose diagonal is just $\rho(x_s)$ (cf. [St2] 2.13, Th.1). We then have the following corollary of Theorem 3.2:

Lemma ([St2] 3.4, Cor. 1, Cor. 4): For all $x \in G$, $f \in k[G]^G$, we have $f(x) = f(x_s)$. In particular, $x \in G$ is unipotent exactly when $f(x) = f(e)$ for all $f \in k[G]^G$.

All unipotent elements of G lie, therefore, in a subvariety $V(G)$ defined by the equations $f(x) - f(e) = 0$, $f \in k[G]^G$. An equivalent definition is that $V(G)$ is the fiber $\chi^{-1}(\bar{e})$ of the adjoint quotient $\chi : G \to T/W$.

Since we do not know from the preceeding whether this variety is reduced (see however 3.10), we denote by V its set of k-rational points $V(G)(k) = V(G)_{red}(k)$. However, both will be called the unipotent variety of G.

If R is a reductive group with semisimple commutator $G = (R,R)$, then all unipotent elements of R will lie in G. We therefore define the unipotent variety $V(R)$ of R as that of G. If $\pi : G \to G'$ is a central separable isogeny of semisimple groups, then π induces an isomorphism of the (reduced) unipotent varieties of G and G', since the kernel of π consists of semisimple elements. The unipotent variety of a product of semisimple groups is isomorphic to the product of the individual unipotent varieties $V(G \times G') = V(G) \times V(G')$.

Remark: An inseparable isogeny $\pi : G \to G'$ may change the type of the unipotent variety. Let for example $char(k) = 2$. The unipotent variety of SL_2 is then given by one equation

$$X^2 + YZ = 0$$

in the three variables X,Y,Z, whereas that of PGL_2 is given by three equations

$$\begin{aligned} X^2 + YZ &= 0 \\ Y(X+1) &= S^2 \\ Z(X+1) &= T^2 \end{aligned}$$

in five variables defining a nonreduced scheme whose underlying reduced variety is nonsingular.

3.10. Fibers of the adjoint quotient. Let G be semisimple of rank r and T be a maximal torus of G. In 3.9 we defined the unipotent variety of G as a special fiber of the adjoint quotient $\chi : G \to T/W$. We now come to the description of the other fibers.

According to [St2] 2.13 Cor. 4, all unipotent elements of the centralizer $Z_G(t)$ of an element $t \in T$ lie in the identity component $Z_G(t)^\circ$. We therefore set $V(Z_G(t)) = V(Z_G(t)^\circ)$.

Lemma: Let $t \in T$ and $V(t)$ be the reduced unipotent variety of the centralizer $Z(t)$ of t in G. Then the morphism $\alpha_t : G \times^{Z(t)} V(t) \to G$, $g * u \mapsto {}^g(tu)$, is a G-equivariant isomorphism onto the reduced fiber $G_{\bar{t}} \cong \chi^{-1}(\bar{t})_{red}$.

Proof: We will construct a G-morphism $\phi : G_{\bar{t}} \to G/Z(t)$ whose fiber $\phi^{-1}(Z(t))$ is just $tV(t)$. The statement will then follow from 3.7 Lemma 4 because $tV(t)$ and $V(t)$ are $Z(t)$-isomorphic.

So let $x \in G_{\bar{t}}$. The semisimple part x_s of x will also be in $G_{\bar{t}}$ because $\chi(x) = \chi(x_s)$. It follows from [St2] 3.4 Cor. 3 that all semisimple elements in $G_{\bar{t}}$ are conjugate to each other and in particular to t. Using the G-isomorphism $gZ(t) \mapsto {}^g t$ we can identify the homogeneous space $G/Z(t)$ with the conjugacy class $C(t)$ of t (cf. [Bo] III. 9.1) and define

$$\phi : G_{\bar{t}} \to C(t) = G/Z(t)$$

by

$$\phi(x) = x_s, \text{ for } x \in G_{\bar{t}}.$$

The semisimple part of ${}^g x$ is equal to ${}^g(x_s)$ by the uniqueness of the Jordan decomposition, which shows that ϕ is G-equivariant. The fiber $\phi^{-1}(t)$ consists of exactly those elements $x \in G_{\bar{t}}$ which have semisimple part t, i.e. which have the form $x = tu$ with $u \in V(G) \cap Z(t) = V(t)$. It remains to be shown that the map ϕ is a morphism.

Recalling the discussion in 3.3 on the Jordan decomposition, it is enough to show that all elements x in $G_{\bar{t}}$ have the same characteristic polynomial for any faithful linear representation ρ of G. This is in fact the case since the characteristic polynomial of $\rho(x)$ is equal to that of the semisimple part $\rho(x_s)$ and for all $x \in G_{\bar{t}}$, the x_s's are conjugate to each other.

Remark: This lemma can be interpreted from the viewpoint of Luna's theory of "ètale slices" (cf. [Lu]). Namely, in a neighborhood of t, $Z(t)$ gives a transverse slice to the closed orbit of t on which the stabilizer $Z(t)$ acts by means of the adjoint action. Given the assumption $\mathrm{char}(k) = 0$ the lemma above is a corollary of [Lu] III.1, Th., Remarques 1.2.

Theorem ([St2] 3.8 Th. 1, Th. 7, [Lus], [Ri]): *Let G be semisimple of rank r and $\chi : G \to T/W$ be the adjoint quotient for G. For a reduced fiber $G_{\bar{t}} = \chi^{-1}(\bar{t})_{red}$ of χ the following holds:*

i) *$G_{\bar{t}}$ is the union of finitely many conjugacy classes.*

ii) *$G_{\bar{t}}$ contains exactly one class of regular elements, which is open and dense in $G_{\bar{t}}$, and whose complement has codimension ≥ 2 in $G_{\bar{t}}$.*

iii) *$G_{\bar{t}}$ contains exactly one class of semisimple elements, the orbit of t, which is the only closed class in $G_{\bar{t}}$, and which lies in the closure of every other class in $G_{\bar{t}}$.*

iv) *$G_{\bar{t}}$ is irreducible and has codimension r in G.*

v) *If G is simply connected the morphism χ is flat and its schematic fibers are reduced and normal.*

vi) *If $\mathrm{char}(k)$ does not divide the order of $\pi_1(G)$ the nonsingular points of $G_{\bar{t}}$ are exactly the regular elements.*

Remark (on the proof of the theorem): Points i) through iv) are found in [St2]

3.8 Th. 1, however, i) also contains the result of Lusztig ([Lus]) on the finiteness of the unipotent conjugacy classes in any characteristic. If char(k) is good (cf. 3.13) Richardson's proof ([Ri], [St2] 3.6) is much more elementary. Another finiteness proof in sufficiently high or zero characteristic follows from Dynkin's theory (cf. 7.3). Our foregoing lemma enables us to reduce the first four statements to corresponding statements about the unipotent varieties. This reduction appears implicitly in the proof of Th. 1 in [St2] 3.8. The assumption that $\pi_1(G) = 0$ which is made there is not required by the proof. Statement v) is proved in [St2] 3.8, Th. 7. The general results from algebraic geometry referred to in [St2] may be found for example in [Ko2] 1.6, Lemma 4 and EGA IV 5.8.6, and the fact that χ is flat is only a reformulation of Steinberg's statement that all fibers of χ are complete intersections (cf. EGA 0_{IV} 15.1.21, IV 19.3). Statement vi) is proved in [St2] 3.8, Th. 7 for simply connected groups (it follows from v) and the differential criterion for regularity (cf. 3.8)). Our more general formulation is derived as follows: If char(k) does not divide the order of $\pi_1(G)$ we obtain G as the quotient $\tilde{G}/C$ (in the sense of [Bo] II. 6) of its universal covering $\tilde{G}$ by a finite central subgroup C isomorphic to $\pi_1(G)$. Let $\tilde{T}$ and $T = \tilde{T}/C$ be maximal tori in $\tilde{G}$ and G. As the translation by C on $\tilde{G}$ resp. $\tilde{T}$ commutes with conjugation by elements of $\tilde{G}$ resp. $N_{\tilde{G}}(\tilde{T})$ the adjoint quotient for G is the quotient by C of that for $\tilde{G}$:

$$\begin{array}{ccc} \tilde{G} & \xrightarrow{/C} & G \\ {\scriptstyle\tilde{\chi}}\downarrow & & \downarrow{\scriptstyle\chi} \\ \tilde{T}/W & \xrightarrow{/C} & T/W \end{array}$$

Hence the fibers of $\tilde{\chi}$ are mapped êtale onto fibers of χ. As the centralizer-dimensions are thereby respected our claim follows.

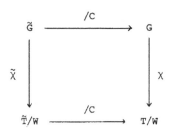

Now let G be semisimple and simply connected. If $t \in T$ is a regular element of a maximal torus T of G, then the centralizer $Z_G(t)$ is equal to T (cf. 3.4, 3.8) and $\chi^{-1}(\bar{t})$ is smooth (3.10) and isomorphic to G/T. Singularities appear in those fibers which are not smooth and so contain irregular elements (3.8). This is exactly the case for fibers $\chi^{-1}(\bar{t})$ with t irregular. According to the description in 3.8, the set of irregular elements in T forms a hypersurface $T_\Sigma = \bigcup_{\alpha \in \Sigma} T_\alpha$, $T_\alpha = \{t \in T | \alpha(t) = 1\}$, whose image under the projection $T \to T/W$ is a closed set $D_{T/W}$ in T/W (EGA II 6.1.10). By the discussion above, $D_{T/W}$ is just the set of critical values of χ, and we call $D_{T/W}$ with its reduced structure the <u>discriminant of</u> χ. Since G is simply connected, all isotropy groups of the W-action on T are reflection groups and the set of branching points of the covering map $T \to T/W$ is exactly T_Σ ([S-S] II.4.2). Therefore $D_{T/W}$ is the discriminant of this covering map.

If G is not simply connected, there can be isotropy groups of the W-action on T which are not reflection groups ([S-S] II.4.4, 4.5, 4.7), and these are responsible for singularities in T/W (cf. also 4.5).

<u>3.11 An Example</u>. It is time to illustrate the foregoing general definitions and results by means of an example. We consider $G = SL_n$. Its Lie algebra $\underline{g} = sl_n$ consists of the trace zero $n \times n$-matrices. The adjoint representation is given by conjugation

$$SL_n \times sl_n \longrightarrow sl_n , \quad (g,x) \longmapsto g\,x\,g^{-1} .$$

As a maximal torus T in SL_n we choose the diagonal matrices

$$T = \left\{ t = \begin{pmatrix} t_1 & & 0 \\ & \ddots & \\ 0 & & t_n \end{pmatrix} \middle| \prod_{i=1}^{n} t_i = 1 \right\} .$$

Its normalizer $N(T)$ in SL_n consists of the monomial matrices

$$N(T) = \{ ((a_{ij})) \in SL_n \mid \exists \sigma \in \mathfrak{S}_n \text{ s.t. } a_{ij} = 0 \text{ for } j \neq \sigma(i) \}$$

whose quotient $W = N(T)/T$ is isomorphic to the symmetric group $\mathfrak{S}_n$ operating on T by permuting the entries.

We denote the character $T \to G_m$, $t \mapsto t_i$, of T by ε_i. The group $X^*(T)$ of all characters may then be identified with the sublattice of $\bigoplus_{i=1}^{n} \mathbb{Z}\,\varepsilon_i = \left\{ \sum_{i=1}^{n} c_i \varepsilon_i \middle| c_i \in \mathbb{Z} \right\}$ defined by $\sum_{i=1}^{n} c_i = 0$. The standard quadratic form $\sum_{i=1}^{n} c_i \varepsilon_i \mapsto \sum_{i=1}^{n} c_i^2$ induces a W-invariant scalar product on $X^*(T)$.

With respect to T the adjoint representation decomposes into eigenspaces

$$sl_n = \text{Lie } T \oplus \bigoplus_{i \neq j} (sl_n)_{ij} .$$

Here $(sl_n)_{ij}$ denotes all matrices with possible non-zero entry only at position ij, and T acts on it by the character $\varepsilon_i - \varepsilon_j : T \to G_m$, $t \mapsto t_i t_j^{-1}$. The corresponding one-parameter subgroup U_{ij} in SL_n is given by $U_{ij} = 1 + (sl_n)_{ij}$ where 1 is the identity matrix and addition that of matrices. All roots of T in sl_n are now given by

$$\Sigma = \{ \varepsilon_i - \varepsilon_j \mid i,j \in \{1,\ldots,n\},\ i \neq j \} .$$

If we choose the upper triangular matrices in SL_n as a Borel subgroup B containing T the corresponding system of simple roots will be

$$\Delta = \{ \alpha_i := \varepsilon_i - \varepsilon_{i+1} \mid i = 1,\ldots,n-1 \}$$

with Dynkin diagram A_{n-1}.

The characters $\lambda_i = \varepsilon_1 + \ldots + \varepsilon_i$, $i=1,\ldots,n-1$, define fundamental dominant weights in $X^*(T)$. This implies that SL_n is simply connected. The irreducible representation

$\rho_{\lambda_i} : SL_n \to GL(V_{\lambda_i})$ with heighest weight λ_i is realized by the i-th exterior power $\Lambda^i\rho$ of the natural representation $\rho : SL_n \to GL(k^n)$. Its trace $\chi_i : SL_n \to k$ appears as coefficient in the characeristic polynomial

$$\det(\xi - x) = \sum_{i=o}^{n} (-1)^i \chi_i(x)\xi^{n-i} \in k[\xi]$$

where $x \in SL_n$. We automatically have $\chi_o(x) = 1$ and $\chi_n(x) = \det(x) = 1$. The adjoint quotient $\chi : SL_n \to T/W$ may thus be viewed as mapping an element $x \in SL_n$ to its characteristic polynomial $\chi(x) = \det(\xi - x)$.

Let $p = \sum_{i=o}^{n} (-1)^i p_i \xi^{n-i}$ be a polynomial in $k[\xi]$ with $p_o = p_n = 1$ corresponding to a point $p \in T/W \cong k^{n-1}$. The multiplicities $m_1(p),\ldots,m_\mu(p)$ of the roots of p determine a partition $m(p)$ of n , i.e. we have $\sum_{i=1}^{\mu} m_i(p) = n$. Up to conjugacy an element x of the fiber $\chi^{-1}(p)$ is determined by a sequence $s_1(x),\ldots,s_\mu(x)$ of partitions $s_i(x) = (s_{i1}(x),\ldots,s_{ie_i}(x))$ of $m_i(p)$ (i.e. $\sum_{j=1}^{e_i} s_{ij}(x) = m_i(p))$, where $s_{ij}(x)$ denotes the size of the j-th block for the i-th eigenvalue t_i of multiplicity m_i in the Jordan normal form of x :

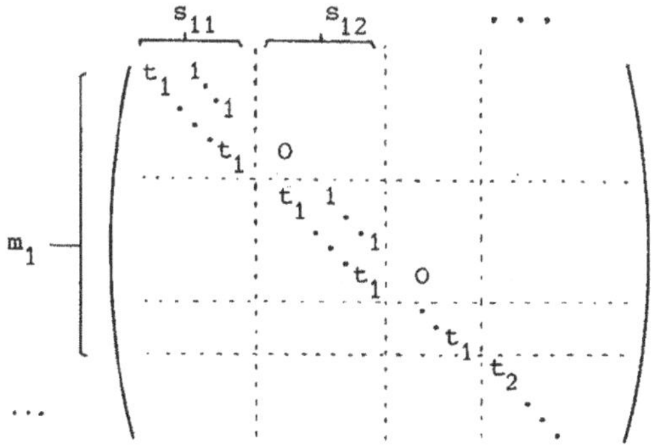

Thus any fiber $\chi^{-1}(p)$ contains only finitely many orbits. An element $x \in \chi^{-1}(p)$ is semisimple exactly when $s_i(x)=(1,1,\ldots,1)$ for all i , the semisimple commutator of $Z_{SL_n}(x)$ being isomorphic to $SL_{m_1(p)} \times \ldots \times SL_{m_\mu(p)}$. An element $x \in \chi^{-1}(p)$ is regular if and only if $s_i(x) = (m_i(p))$ for all i . The unipotent variety is given by $\chi^{-1}(\chi(1))$, and $\chi(1) = (\xi - 1)^n$. Its only semisimple element is $\begin{pmatrix} 1 & & 0 \\ & \ddots & \\ 0 & & 1 \end{pmatrix}$,

and its regular elements are conjugate to $\begin{pmatrix} 1 & 1 & & \\ & \ddots & \ddots & \\ & & \ddots & 1 \\ & & & 1 \end{pmatrix}$.

Exactly when $m(p)=(1,1,\ldots,1)$ the fiber $\chi^{-1}(p)$ is nonsingular consisting of only one orbit of regular semisimple elements. The discriminant $D_{T/W}$ in T/W corresponds thus to the set of characteristic polynomials p for which the usual discriminant vanishes.

3.12 Invariants on the Lie Algebra. Instead of looking at the quotient $\chi : G \to T/W$ we can consider the quotient of the adjoint representation $\gamma : \underline{g} \to \operatorname{Spec} k[\underline{g}]^G$. For $k = \mathbb{C}$ these investigations were done by Kostant in [Ko2] (before those of Steinberg [St1]) and finally extended by Veldkamp [Ve] to the case when char(k) doesn't divide the order of the Weyl group.

Let $\underline{h}$ be the Lie algebra of a maximal torus T of the semisimple group G . Then the Weyl group $W = N(T)/T$ operates in a natural way on $\underline{h}$. Let $k[\underline{h}]^W$ be the W-invariant polynomials on $\underline{h}$. An analogous result to Theorem 3.2 will hold, which goes back to Chevalley for the case when $\operatorname{char}(k) = 0$.

Theorem ([S-S] II 3.17', [De1] Th. 3, Cor., LIE VIII, § 8, n° 3, Th. 1, Cor. 1): Let the semisimple group G be adjoint. Then the inclusion $\underline{h} \subset \underline{g}$ induces an isomorphism $k[\underline{g}]^G \to k[\underline{h}]^W$. If G is also simple of rank r and char(k) is not a torsion prime for G , then there exist r algebraically independent, homogeneous polynomials $\gamma_1,\ldots,\gamma_r \in k[\underline{g}]^G$ of degrees $m_1+1,\ldots,m_r+1$, where $m_1,\ldots,m_r$ are the exponents of W , which generate the k-algebra $k[\underline{g}]^G$.

For a definition of the exponents of W see LIE V, § 6, n° 2, and for their values, LIE IV, V, VI, Planches I - IX (cf. also 7.4).

Remark: The somewhat more general formulation of the above theorem (than that found in the literature) is obtained by the following arguments. Let $\pi : \tilde{G} \to G$ be the simply connected cover of G . If char(k) is not a torsion prime for G , then π is separable, that is $d\pi : \mathrm{Lie}(\tilde{G}) \to \mathrm{Lie}(G)$ is an isomorphism (cf. 3.6). Because $d\pi \circ \mathrm{Ad}_{\tilde{G}}(g) = \mathrm{Ad}_G(\pi(g)) \circ d\pi$ for all $g \in \tilde{G}$ (cf. [Bo] I 3.5(3)), the adjoint representation of $\tilde{G}$ on $\mathrm{Lie}(\tilde{G})$ factors over that of G on $\mathrm{Lie}(G) = \underline{g}$. Therefore [De1] Th. 3, Cor. can also be used for the adjoint group G . The degrees of the γ_i are uniquely determined by the formula in [De1] Th. 3, and hence are the same as in the case when char(k) = 0 (cf. LIE VIII § 8, n° 3, Th. 1, Cor. 1). As the above argument shows, the assumption that G is adjoint is superfluous when char(k) doesn't divide the order of $L^*(\Sigma)/L(\Sigma)$ (cf. 3.6).

The morphism induced by the inclusion $k[\underline{h}]^W \to k[\underline{g}]$ will be called the <u>adjoint quotient of</u> $\underline{g}$ and will be written simply $\gamma : \underline{g} \to \underline{h}/W$ (cf. also 3.2). The class of $h \in \underline{h}$ in $\underline{h}/W$ is denoted by $\bar{h}$. A fiber of $\gamma_{|\underline{h}} : \underline{h} \to \underline{h}/W$ will consist of exactly one W-orbit for reasons analogous to those in 3.2.

The investigation of the fibers of γ requires further restrictions on the characteristic of k to 0 or "very good" primes.

<u>3.13 Very Good Primes.</u> (This section will be needed only in the case of char(k) = p > 0.) Let G be a semisimple group with maximal torus T and root system $\Sigma \subset X^*(T) \otimes_{\mathbb{Z}} \mathbb{Q}$. If Σ' is a subset of Σ , let $L(\Sigma')$ be the $\mathbb{Z}$-lattice generated in $X^*(T)_{\mathbb{Q}}$ by Σ' (cf. 3.6).

<u>Definition:</u> A prime p is called <u>good</u> (resp. <u>bad</u>) for Σ if there does not exist (resp. there does exist) a $\mathbb{Z}$-closed root subsystem Σ' of Σ such that the quotient $L(\Sigma)/L(\Sigma')$ has torsion of order p .

To determine the bad primes, it obviously suffices to look at the irreducible root systems.

<u>Lemma</u> ([S-S] I 4.3, 4.4): <u>A prime</u> p <u>is bad for an irreducible root system</u> Σ <u>exactly when</u> p <u>is a torsion prime of</u> Σ , <u>or is equal to</u> 2 <u>for</u> $\Sigma = C_r$, <u>or is</u>

<u>equal to</u> 3 <u>for</u> $\Sigma = G_2$. <u>If</u> p <u>is good for</u> Σ <u>it will also be good for all root subsystems</u> Σ' <u>of</u> Σ .

<u>Definition</u>: The characteristic p of k is called <u>good</u> for G when $p = 0$ or p is good for Σ . If p is good and does not divide $r + 1$, when Σ contains a component of type A_r , it is called <u>very good</u>.

When the characteristic is very good, all central isogenies in the central isogeny class of G are separable, in particular the isomorphism type of the Lie algebra is independent of any representative of this class and all adjoint representations factor over that of the adjoint group (cf. also 3.6 and Remark 3.12).

If the characteristic of k is good for G , this has important consequences for the investigations in chapter 5. Let $G \to \operatorname{Aut} G$ resp. $G \to \operatorname{Aut}(\underline{g})$ be the adjoint action resp. representation of G . According to [Bo] III 9.1, the orbit maps $G \to G$, $g \mapsto gxg^{-1}$ resp. $G \to \underline{g}$, $g \mapsto gyg^{-1}$ are separable morphisms for all $x \in G$ resp. $y \in \underline{g}$ exactly when the global and infinitésimal centralizers have the same dimension, i.e. when

$$\operatorname{Lie}(Z_G(x)) = \underline{z}_{\underline{g}}(x) = \{z \in \underline{g} \mid \operatorname{Ad}(x)z = z\} \quad \text{for all} \quad x \in G$$

resp.

$$\operatorname{Lie}(Z_G(y)) = \underline{z}_{\underline{g}}(y) = \{z \in \underline{g} \mid [z,y] = 0\} \quad \text{for all} \quad y \in \underline{g} \ .$$

<u>Theorem</u> (Richardson, [S-S] I 5.1 - 5.6): <u>Let</u> G <u>be either equal to</u> GL_n <u>or simple and</u> char(k) <u>very good for</u> G . <u>Then the orbit maps of the adjoint action and representation are all separable</u>.

<u>Remark 1</u>: The somewhat more general formulation of the result than that in [S-S] can be gotten in a trivial way: when the characteristic is very good, it suffices to prove the statement for a representative of the central isogeny class of G (see above), and in the case A_{n-1} , $p \nmid n$, an SL_n-stable complement to sl_n consists of the scalar matrices in gl_n .

Remark 2: If char(k) is not very good for G, the theorem will be false, for example, in the case A_{n-1}, $p \mid n$, the morphism $SL_n \to PGL_n$ is inseparable. For other examples see [Sp1] 5.9.

3.14 The Fibers of γ. Let $G, T, W, \underline{g}, \underline{h}$ be as in 3.12. We can investigate the fibers of $\gamma : \underline{g} \to \underline{h}/W$ with a similar approach to the way we investigated the fibers of $\chi : G \to T/W$. We will assume in this section that the characteristic of k is very good for G.

The roots of $\underline{h}$ in $\underline{g}$. If $X^*(T)$ is the character group of T, then the map $X^*(T) \to \underline{h}^*$, which sends a character $\omega : T \to G_m$ to its differential $D_e\omega : T_e(T) = \underline{h} \to T_eG_m = k$, induces a natural isomorphism of $X^*(T) \otimes_{\mathbb{Z}} k$ with $\underline{h}^*$. If the characteristic of k is very good (as assumed), then this map is injective on the root system $\Sigma \subset X^*(T)$ of T in G, and the image of Σ will consist of just those linear forms $\alpha \in \underline{h}^* - \{0\}$ for which the eigenspace $\underline{g}_\alpha = \{x \in \underline{g} \mid [h,x] = \alpha(h)x \text{ for all } h \in \underline{h}\}$ is nontrivial. So we will call the image of Σ in $\underline{h}^*$ the roots of $\underline{h}$ in $\underline{g}$, and denote it by Σ, too.

The Jordan decomposition in $\underline{g}$. Analogous to the multiplicative Jordan decomposition in G there is an additive Jordan decomposition $x = x_s + x_n$ of an element $x \in \underline{g}$ into a semisimple part x_s, and a nilpotent part x_n, which commute with each other and are uniquely determined. Semisimplicity and nilpotence in $\underline{g}$ can be defined effectively by the differential $D_e\rho : \underline{g} \to gl(V)$ of any faithful linear representation $\rho : G \to GL(V)$. The Jordan decomposition of $x \in \underline{g}$ can then be gotten from the decomposition of $D_e\rho(x) \in gl(V) = End(V)$ (cf. [Bo] I 4.4).

Centralization of semisimple elements. Due to the assumption on char(k), the centralizer $Z_G(h)$ of every semisimple element $h \in \underline{g}$ is reductive and connected ([St3] 3.14). If we interpret h without loss of generality as an element of $\underline{h}$ (cf. [Bo] IV 11.3, 11.8), we can identify the root system Σ_h of T in $Z_G(h)$ with those roots $\alpha \in \Sigma$ of T in G which as elements in $\underline{h}^*$ vanish at h (loc.cit., this also follows from the fact that $\mathrm{Lie}(Z_G(h)) = \underline{z}_{\underline{g}}(h))$. From the k-linearity of the equation $\alpha(h) = 0$ for $\alpha \in \Sigma_h$, and the fact that char(k) is good for Σ , it follows easily that Σ_h is a $\mathbb{Q}$-closed root subsystem of Σ .

<u>Regularity</u>. An element $x \in \underline{g}$ is called <u>regular</u> when its G-orbit in $\underline{g}$ has maximal dimension. This is equivalent to the global (or infinitesimal) centralizer of x having minimal dimension. As in the case of the adjoint action of G , it is easy to see that the minimal dimension of a centralizer is equal to the rank of G .

Replacing "unipotent" by "nilpotent" and the roots of T in G by the roots of $\underline{h}$ in $\underline{g}$ in 3.8, we get analogous characterization and existence theorems for regular elements in $\underline{g}$ (cf. [Ve] 4.6, 4.7, [Sp1]).

<u>The fibers of</u> γ . Given the quotient $\gamma : \underline{g} \to \underline{h}/W$, the equation $\gamma(x) = \gamma(x_s)$ will hold for all $x \in \underline{g}$. In particular, $\gamma^{-1}(\bar{0})$ consists of the nilpotent elements of $\underline{g}$. Therefore we call $\gamma^{-1}(\bar{0})$ the <u>nilpotent variety</u> $N(\underline{g})$ of $\underline{g}$ (which is schematically reduced, see below). The statements of 3.10 Lemma and Theorem corresponding to a simply connected group G will hold for all fibers of γ (cf. [Ve] Part III, the stronger assumptions there on char(k) are unnecessary because of the results of Demazure [De1]).

Moreover, the nilpotent variety $N(\underline{g})$ is G-isomorphic to the unipotent variety V(G) of G . In char(k) = 0 , this follows from 3.15. In char(k) = p > 0 , p very good for G , this follows from [Sp2] 3. and the normality of $N(\underline{g})$ ([Ve] 6.9).

<u>The discriminant of</u> γ . Let $\underline{h}_\Sigma \subset \underline{h}$ be the union of the root hypersurfaces $\underline{h}_\alpha = \{h \in \underline{h} \mid \alpha(h) = 0\}$. Under the projection $\underline{h} \to \underline{h}/W$ the set $\underline{h}_\Sigma$ will be mapped to a closed set $D_{\underline{h}/W}$ in $\underline{h}/W$, over which the singular fibers of γ will lie by an argument similar to that in 3.10. The set $D_{\underline{h}/W}$ with the reduced structure will be called the <u>discriminant</u> of γ . In addition, the preimage of the branch locus of the branched covering $\underline{h} \to \underline{h}/W$ is $\underline{h}_\Sigma \subset \underline{h}$, so that $D_{\underline{h}/W}$ is also the (reduced) discriminant of that cover (cf. LIE V § 5, Prop. 6).

In the following section we will cite a result that (at least for $\mathrm{char}(k) = 0$) permits a direct passage from statements about $\chi : G \to T/W$ to statements about $\gamma : \underline{g} \to \underline{h}/W$.

3.15 A Comparison Theorem. In this section, let G be a semisimple group and $\mathrm{char}(k) = 0$. As a special case of more general results of Luna on the actions of reductive groups on affine varieties in $\mathrm{char}(k) = 0$, we get the following relation between the quotients $\chi : G \to T/W$ and $\gamma : \underline{g} \to \underline{h}/W$.

Theorem: There are affine open neighborhoods U of $\bar{e}$ in T/W and U' of $\bar{0}$ in $\underline{h}/W$, a surjective étale morphism $\pi/G : U \to U'$ with $\pi/G(\bar{e}) = \bar{0}$, and a G-isomorphism $\chi^{-1}(U) \tilde{\to} \gamma^{-1}(U') \times_{U'} U$. In particular, for all $u \in U$ the fibers $\chi^{-1}(u)$ and $\gamma^{-1}(\pi/G(u))$ are isomorphic as G-varieties.

Proof: Apply [Lu] III.1, Th. Remarque 1, and II.2, Lemma 3 to the case $X = G$, $x = e \in G$, $T_eX = \underline{g}$, $G_e = Z_G(e) = G$.

Remarks: i) The result that $N(\underline{g})$ is isomorphic with $V(G)$, the unipotent variety of G , follows from the theorem since $\pi/G(\bar{e}) = \bar{0}$.

ii) Without some further restrictions, the theorem can be false in the case of positive characteristic, as the example $G = SL_2$ with $\mathrm{char}(k) = 2$ shows. In that case, $\chi : SL_2 \to T/W$ and $\gamma : sl_2 \to \underline{h}/W$ give two nonequivalent deformations of the ordinary double point in the fiber $\chi^{-1}(\bar{e}) \cong \gamma^{-1}(\bar{0})$.

iii) From the viewpoint of the étale topology, the theorem gives an isomorphism of χ and γ over étale neighborhoods of $\bar{e} \in T/W$ and $\bar{0} \in \underline{h}/W$. In particular, in the case $k = \mathbb{C}$ the étale morphism $\pi/G : U \to U'$ can be replaced by an analytic isomorphism of neighborhoods in the usual Hausdorff topology. The theorem can then be proved by means of the exponential map $\underline{g} \to G$.

iv) The theorem allows the reduction of statements about the fibers of γ to statements about the fibers of χ , but not the reverse. Since the invariant polynomials of $\underline{g}$ can be generated by homogeneous ones, γ will be a G_m-equivariant morphism, when $\underline{g}$ is given the usual scalar action of G_m, and $\underline{h}/W$ the G_m-action defined by the degrees of the homogeneous generators of $k[\underline{h}]^W$. Since the degrees are positive, every fiber of γ can be translated by G_m into one over any small neighborhood of $\bar{0} \in \underline{h}/W$. On the other hand, the torus T of G contains elements t for which the root system $\Sigma_t = \{\alpha \in \Sigma \mid \alpha(t) = 1\}$ is not $\mathbb{Q}$-closed. These elements lie outside the neighborhood Q of $e \in T$ defined in 3.5, and the isomorphism type of the corresponding fibers $\chi^{-1}(\bar{t})$ can not appear over some small neighborhood of $\bar{e} \in T/W$. Differences also appear in a consideration of the quotients T/W and $\underline{h}/W$. If G is not simply connected, T/W can possess singularities (cf. 4.5), while $\underline{h}/W$ is always an affine space (cf. LIE V § 5, n° 5.3, Th. 3). It does follow from the theorem that T/W is smooth at $\bar{e}$ (π/G étale).

4. The Resolution of the Adjoint Quotient

4.1. The Resolution of the Singularities of the Unipotent Variety.

Let G be a semisimple group and $\chi : G \to T/W$ its adjoint quotient. If char(k) does not divide the order of the fundamental group of G, then the singular points of the reduced fibers of χ correspond to the irregular elements of G (cf. 3.10). A resolution is useful in the investigation of these singularities.

<u>Definition</u> ([Hi] intro.): A morphism $\pi : Y \to X$ of reduced varieties is a <u>resolution of the singularities of</u> X, when the following conditions hold:

i) π is proper,

ii) Y is smooth,

iii) the preimage $\pi^{-1}(X^{reg})$ of the smooth points of X is a dense set in Y, and π induces an isomorphism $\pi^{-1}(X^{reg}) \xrightarrow{\sim} X^{reg}$.

Now let G be reductive and B a Borel subgroup of G. Since all Borel subgroups of G are conjugate to B, and B is its own normalizer ([Bo] 11.1, 11.15), we may identify the set $\mathbb{B}$ of all Borel subgroups of G with the complete variety G/B by means of the bijection $G/B \to \mathbb{B}$, $gB \mapsto gBg^{-1}$.

Let $U \subset B$ be the unipotent radical of B, which as a normal subgroup will be stable under the adjoint action of B on itself. The associated bundle $G \times {}^B U$ is then a smooth irreducible variety of dimension $\dim G - \operatorname{rank} G$.

The group U lies in the unipotent variety V of G, on which G operates by the adjoint action. We therefore can define a closed embedding of $G \times {}^B U$ in $G/B \times V$ by the morphism $\tau : G \times {}^B U \to G/B \times V$, $\tau(g * u) = (gB, {}^g u)$, cf. 3.7. Identifying $\mathbb{B}$ with G/B we can describe the image under τ as the set of all pairs $(A,x) \in \mathbb{B} \times V$ with $x \in A$.

<u>Theorem</u> ([Sp2] 1.4, [St4] 1.1): <u>Let</u> G <u>be reductive where</u> char(k) <u>does not divide the order of the fundamental group of</u> G (cf. 3.6). <u>Then the morphism</u>

$\pi : G \times {}^B U \to V$, $\pi(g * u) = {}^g u$, is a resolution of the singularities of the unipotent variety V of G .

Proof: The proof of the theorem for $\mathrm{char}(k) = 0$ is very simple. The morphism π factors through the second projection $p : G/B \times V \to V$, $\pi = p \circ \tau$. Therefore π is proper. The variety V is the closure of its regular orbit, which consists of the smooth points of V . Since $G \times {}^B U$ is irreducible and π is G-equivariant, we need only show that the preimage $\pi^{-1}(x)$ of a regular unipotent element $x \in V$ consists of only one point. However, $\tau(\pi^{-1}(x)) = \{(A,x) \in \mathcal{B} \times \{x\} \mid x \in A\}$, and by the characterization in 3.8, x lies in exactly one Borel subgroup $A \in \mathcal{B}$.

If the characteristic of k is positive, the separability of π must be proved. If $\mathrm{char}(k)$ is very good for (G,G) , this follows from the separability of the orbit maps (cf. 3.13). Otherwise, a detailed analysis is necessary, which leads to the stated condition on the fundamental group. For that see [St4] 1.1, 6.1.

Definition: Let the morphism $\pi : Y \to X$ be a resolution of the singularities of X . The preimage $\pi^{-1}(X^{\mathrm{sing}})$ of the singular points of X endowed with the reduced structure is called the reduced exceptional set $E(\pi)$ of π . If for every point $y \in E(\pi)$ there is a local coordinate system $y_1,\dots,y_n$ such that in a neighborhood of y , $E(\pi)$ is given by an equation of the form $y_1 \cdot \ldots \cdot y_k = 0$, $k \leq n$, then $E(\pi)$ is a divisor with normal crossings.

The reduced exceptional set $E(\pi)$ of the resolution $\pi : G \times {}^B U \to V$ of the theorem above is a divisor with normal crossings. The G-equivariance of π implies that $E(\pi)$ has the form $G \times {}^B(U \cap E(\pi))$ (cf. 3.7 Lemma 4) where $U \cap E(\pi)$ will consist of the irregular elements of U . For $x \in U$, we can represent x in the form $x = \prod_{\alpha \in \Sigma^+} \phi_\alpha(c_\alpha)$ as an element of the product $U = \prod_{\alpha \in \Sigma^+} U_\alpha$ (cf. 3.8). Then the irregular elements of U can be characterized as just those elements for which at least one of the c_α with $\alpha \in \Delta$ (Δ the basis of Σ^+) vanishes. Therefore $U \cap E(\pi)$ is defined as a subset of the affine space U where the product of $r(=\mathrm{rank}\ G)$ coordinate functions vanishes. With that we see that $E(\pi)$ is a divisor with normal crossings in $G \times {}^{B}U$.

<u>4.2. Simultaneous Resolutions.</u> In 4.1 we have constructed a resolution of the singularities of the unipotent variety of a semisimple group. Lemma 3.10 suggests the following for the other fibers of the quotient morphism $\chi : G \to T/W$. Let $t \in T$ be an element of the maximal torus T of G . Then $B(t) = B \cap Z_G(t)$ is connected ([Bo] III 10.6(5)), and is a Borel subgroup of $Z_G(t)^\circ$ with unipotent radical $U(t) = U \cap Z_G(t)$, (cf. 3.4). Let $V(t)$ be the reduced unipotent variety of $Z_G(t)$, (cf. 3.10). The morphism $\pi_t : Z(t)^\circ \times {}^{B(t)}U(t) \to V(t)$ defined in 4.1 is a resolution of the singularities of $V(t)$, provided that $Z(t)^\circ$ has a separable universal cover. If $Z(t)$ is connected, as will be the case for simply connected groups G (cf. 3.4), then the morphism

$$G \times {}^{Z(t)}\pi_t \;:\; G \times {}^{Z(t)}(Z(t) \times {}^{B(t)}U(t)) \to G \times {}^{Z(t)}V(t)$$

will be a resolution of the singularities of the fiber $G_{\bar{t}} = \chi^{-1}(\bar{t})_{red}$.

In the next section we will show how these individual resolutions parametrized by T can be put into an algebraic family. For this we define (cf. [Br1]):

<u>Definition:</u> A <u>simultaneous resolution of a morphism</u> $\chi : X \to S$ of reduced varieties consists of a commutative diagram

$$\begin{array}{ccc} Y & \xrightarrow{\ \phi\ } & X \\ {\scriptstyle\theta}\big\downarrow & & \big\downarrow{\scriptstyle\chi} \\ T & \xrightarrow{\ \psi\ } & S \end{array}$$

of morphisms of reduced varieties with the following properties:

i) θ is smooth,

ii) ψ is finite and surjective,

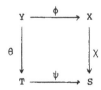

iii) ϕ is proper,

iv) for all $t \in T$, the morphism $\phi_t : \theta^{-1}(t) \to X_{\psi(t)}$ induced by ϕ is a resolution of the singularities of the reduced fiber $\chi^{-1}(\psi(t))_{red} = X_{\psi(t)}$.

4.3 Several Constructions. The simultaneous resolution of the adjoint quotient $\chi : G \to T/W$ of a semisimple group G can be obtained as a special case of a more general construction.

Let G be semisimple, not necessarily simply connected, let B be a Borel subgroup containing a maximal torus T, and let $P \supset B$ be a parabolic subgroup of G ([Bo] IV 11.2). If we let $\Delta \subset \Sigma$ be the system of simple roots of T in G determined by B, P will be uniquely determined by a subset $\Delta_P \subset \Delta$: If Σ_P^+ is the set of those positive roots which are a linear combination of elements of Δ_P, then P will be generated as a group from B and the one-dimensional subgroups $U_{-\alpha}$, $\alpha \in \Sigma_P^+$. Furthermore P can be written as a semidirect product $M \ltimes U$ of the reductive group M generated by T and the U_α, $U_{-\alpha}$, $\alpha \in \Sigma_P^+$, and the unipotent radical U which will be generated by the U_α, $\alpha \in \Sigma^+ \setminus \Sigma_P^+$. Let C be the center $\{t \in T \mid \alpha(t) = 1 \text{ for all } \alpha \in \Delta_P\}$ of M. Then $A = C.U$ is semidirect and $A^\circ = C^\circ.U$ is equal to the solvable radical of P (for details, see [Bo-Ti] 4., [Hu] 30.2). The group P operates on A and U by conjugation.

Lemma 1: P operates trivially on the quotient $A/U \cong C$.

Proof: Since C is commutative, the projection $A \to C$ is invariant with respect to the conjugation on A by elements from $U \subset A$. On the other hand, M centralizes C and normalizes U. Since $P = M \ltimes U$, the projection $A \to C$ is invariant under conjugation on A by elements of P.

Giving C the trivial P-action, the map $A \to C$ will induce a smooth bundle morphism (cf. 3.7)

$$G \times {}^{P}A \to G \times {}^{P}C \xrightarrow{\sim} G/P \times C$$

which together with the projection to the second factor of $G/P \times C$ gives a smooth morphism

$$\theta : G \times {}^{P}A \to C \ .$$

For an element $t.u \in C.U = A$ and $g \in G$ we then have $\theta(g * tu) = t$.

If we embed $G \times {}^{P}A$ as a closed subbundle into $G/P \times G$ as in 3.7, and define $\phi(g * a) := {}^{g}a$ and $p_2(gP,g') = g'$, we get a commutative diagram

$$\begin{array}{ccc} G \times {}^{P}A & \xrightarrow{\ \phi\ } & G \\ & {\scriptstyle \tau} \searrow \quad \nearrow {\scriptstyle p_2} & \\ & G/P \times G & \end{array}$$

where ϕ is a proper morphism (G/P is complete), which in the case $P = B$ is also surjective, since every element in G is conjugate to an element in B ([St2] 2.13, Th. 1).

Let ψ denote the composition $C \hookrightarrow T \to T/W$ of the inclusion of C in T and the finite surjective quotient $T \to T/W$.

Since the image $\chi({}^{g}(tu))$ of $\phi(g * tu)$ is equal to the image $\chi((tu)_s)$ of the semisimple part of tu (cf. 3.9) the following lemma implies the commutativity of the diagram

$$\begin{array}{ccc} G \times {}^{P}A & \xrightarrow{\ \phi\ } & G \\ {\scriptstyle \theta}\downarrow & & \downarrow{\scriptstyle \chi} \\ C & \xrightarrow{\ \psi\ } & T/W \end{array} \ .$$

Lemma 2: Let $t \in C$, $u \in U$. The semisimple part of $tu \in A$ is conjugate under P to t .

Proof: The group C contains only semisimple elements. Therefore the unipotent part v of tu lies in the kernel of the projection $p : A \to C$, und tuv^{-1} is semisimple and equal to $(tu)_s$. Therefore it is enough to show that semisimple elements of the form $tu \in A$ are conjugate to t under P . The fibers of the projection $p : C \ltimes U \to C$ consist of P-conjugacy classes according to Lemma 1. By [Bo] III. 10.6, every semisimple element of $A \subset B$ is conjugate to an element of $T \cap A = C$ in $B \subset P$. However, $p(tu) = t$, and p is injective on C , so tu can only be conjugate to t .

Remark: One can easily show that the projection $A \to C \cong A/U$ is actually the quotient of A by the adjoint action of P . This in turn implies that $\theta : G \times {}^P A \to C$ is the quotient of $G \times {}^P A$ by the natural G-action. The commutativity of the above diagram then simply follows by the G-equivariance of ϕ .

Now let $t \in C$. The fiber $\theta^{-1}(t)$ is by the definition of θ just the subbundle $G \times {}^P tU$ of $G \times {}^P A$. The reduced fiber $G_{\bar{t}} = \chi^{-1}(\bar{t})_{red}$ of the morphism $\chi : G \to T/W$ has been described in 3.10 by means of the isomorphism $\alpha_t : G \times {}^{Z(t)} V(t) \to G_{\bar{t}}$, where $V(t)$ is the reduced unipotent variety of the centralizer $Z(t)$ of t in G . The group $U(t) = U \cap Z(t)$ is the unipotent radical of the parabolic subgroup $P(t) = P \cap Z(t)$ of $Z(t)$. (It follows from [Bo] III 10.6(5) that $U(t)$ is connected; because $P = M \ltimes U$ and $M \subset Z(t)^\circ$ we thus obtain $P(t) = M \ltimes U(t) \subset Z(t)^\circ$; moreover, $P(t)$ contains the Borel subgroup $B(t)$ of $Z(t)$, cf. 4.2.)

Defining a $Z(t)$-morphism

$$\pi_t : Z(t) \times {}^{P(t)} U(t) \to V(t) \quad \text{by} \quad \pi_t(z * u) = {}^z u$$

and a G-morphism

$$\beta_t : G \times^{Z(t)} (Z(t) \times^{P(t)} U(t)) \to G \times^{P} tU \quad \text{by} \quad \beta_t(g * z * u) = gz * tu$$

we have:

<u>Lemma 3:</u> <u>The following diagram is commutative</u>

$$\begin{array}{ccc} G \times^{Z(t)} (Z(t) \times^{P(t)} U(t)) & \xrightarrow{G \times^{Z(t)} \pi_t} & G \times^{Z(t)} V(t) \\ \Big\downarrow \beta_t & & \Big\downarrow \alpha_t \\ G \times^{P} tU & \xrightarrow{\phi_t} & G_t^- \end{array}$$

<u>and</u> β_t <u>is an isomorphism.</u>

<u>Proof:</u> The commutativity holds since $\phi_t \circ \beta_t(g * z * u) = {}^{gz}(tu) = {}^{g}(t\,{}^{z}u) = \alpha_t(g * {}^{z}u) = \alpha_t\ (G \times^{Z(t)} \pi_t)(g * z * u)$.

We factor β_t into the composition of the following G-morphisms

$$G \times^{Z(t)} (Z(t) \times^{P(t)} U(t)) \xrightarrow{\sim} G \times^{P(t)} U(t) \xrightarrow{\sim} G \times^{P} (P \times^{P(t)} U(t)) \to G \times^{P} tU$$

$$g * z * u \quad \mapsto gz * u \quad \mapsto gz * e * u \quad \mapsto gz * tu$$

in which the first two are ismorphisms (3.7 Lemma 2) and the last has the form $G \times^{P} \gamma_t$ where $\gamma_t : P \times^{P(t)} U(t) \to tU$ is defined by $\gamma_t(b * u) = {}^{b}(tu)$. Therefore is suffices to show that γ_t is a P-isomorphism. The proof is analogous to that of Lemma 3.10: we identify the homogeneous space $P/P(t)$ with the P-conjugacy classes $C(t)$ of t by the P-isomorphism $pP(t) \mapsto {}^{p}t$ (cf. [Bo] III.9.1). By 4.3 Lemma 2 the semisimple part x_s of an element $x \in tU$ is conjugate to t under P . Since all elements of tU have the same characteristic polynomial, namely that of

t , for any faithful representation of P , we can define a P-equivariant morphism (cf. 3.3)

$$\omega : t.U \to C(t) = P/P(t) \quad \text{by} \quad \omega(x) = x_s \ .$$

The fiber $\omega^{-1}(t)$ is just $tU(t)$ which is $P(t)$-isomorphic to $U(t)$. The assertion about γ_t follows now from 3.7 lemma 4.

Remark: That $\theta^{-1}(t)$ is a fiber bundle associated to $G \to G/Z(t)$ can also be seen by looking at the G-equivariant composition $\theta^{-1}(t) \to G_{\bar{t}} \to G/Z(t)$. The fiber of this composition is easily identified with $Z(t) \times^{P(t)} U(t)$.

4.4 The Simultaneous Resolution of χ . We now consider the morphisms θ, ϕ, ψ defined in 4.3 in the situation $P = B$, i.e. P is now a Borel subgroup of G . Then we have $P = B$, $A = B$, and $C = T$.

Theorem (Grothendieck): *Let the semisimple group G be simply connected and let char(k) not be a torsion prime for G . Then the following diagram is a simultaneous resolution of the morphism $\chi : G \to T/W$:*

$$\begin{array}{ccc} G \times^B B & \xrightarrow{\ \phi\ } & G \\ \theta \big\downarrow & & \big\downarrow \chi \\ T & \xrightarrow{\ \psi\ } & T/W \end{array}$$

If char(k) is a torsion prime, there still is an open W-stable neighborhood Q of e in T , so that the restricted diagram

$$\begin{array}{ccc} \theta^{-1}(Q) & \longrightarrow & \chi^{-1}(\psi(Q)) \\ \big\downarrow & & \big\downarrow \\ Q & \longrightarrow & \psi(Q) \end{array}$$

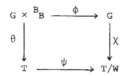

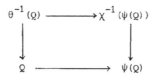

is a simultaneous resolution of χ over $\psi(Q)$.

Proof: Since G is simply connected, all centralizers $Z(t)$, $t \in T$, are connected (3.4). If $\mathrm{char}(k)$ is not (resp. is) a torsion prime for G , then $\mathrm{char}(k)$ does not divide the order of the fundamental group $\pi_1(Z(t))$ for all $t \in T$ (resp. for all t in the open neighborhood Q of e in T as defined in Lemma 3.5) (cf. 3.6). Therefore the morphism $\pi_t : Z(t) \times^{B(t)} U(t) \to V(t)$ is a resolution of the singularities of $V(t)$ for all $t \in T$ (resp. $t \in Q$) (cf. 4.2). Using 4.3 Lemma 3 we see that property iv) of a simultaneous resolution is fulfilled for the above diagrams. The other properties being established in 4.3 the proof is thus completed.

4.5 The Situation for Non-simply Connected Groups. (This section is a digression and will not be used anywhere else.) Now let G be semisimple, but not simply connected, and suppose $\mathrm{char}(k)$ is not a torsion prime. Then G is the quotient (in the sense of [Bo] II.6) of the universal cover $\tilde{G}$ of G by a finite central subgroup $C \subset \tilde{G}$. Let $\tilde{T}$ and $\tilde{B}$, $\tilde{T} \subset \tilde{B}$, be a maximal torus and a Borel subgroup of $\tilde{G}$. Then they contain the center of $\tilde{G}$, in particular they contain C , and their images T and B in G are again a maximal torus and a Borel subgroup.

According to [S-S] II.4.5 there is an element $t \in T$ whose centralizer is not connected. Let t be one such element. By repeated applications of 3.7 Lemma 2 we get a G-isomorphism in the top line of the commutative diagram

$$\begin{array}{ccc} G \times^{Z(t)^\circ} (Z(t)^\circ \times^{B(t)} U(t)) & \xrightarrow{\sim} & G \times^{Z(t)} (Z(t) \times^{B(t)} U(t)) \\ \downarrow & & \downarrow \\ G/Z(t)^\circ & \longrightarrow & G/Z(t) \end{array}$$

in which all the other morphisms are the canonical ones. This shows that the morphism $G \times^{Z(t)} \pi_t$ of 4.3 Lemma 3 (and correspondingly ϕ_t) factors into a resolution of the variety $G \times^{Z(t)^\circ} V(t)$ and an n-sheeted covering, where n is the order of $Z(t)/Z(t)^\circ$,

$$\begin{array}{ccccc} G \times^{Z(t)^\circ} (Z(t)^\circ \times^{B(t)} U(t)) & \longrightarrow & G \times^{Z(t)^\circ} V(t) & \rightarrow & G \times^{Z(t)} V(t) \\ \downarrow & & \downarrow & & \downarrow \\ G/Z(t)^\circ & \longrightarrow & G/Z(t)^\circ & \longrightarrow & G/Z(t) \end{array}$$

In contrast to the variety $V(t)$, on which both $Z(t)^\circ$ and $Z(t)$ operate, there is no apparent extension of the $Z(t)^\circ$-action to a $Z(t)$-action on the resolution $Z(t)^\circ \times^{B(t)} U(t)$, which a natural definition of a resolution of $G \times^{Z(t)} V(t)$ would have made possible.

The fibers of the morphism θ fall, as it were, into the simply connected case, whereas the fibers of χ do not. This can be stated more precisely as follows.

The action of C on $\tilde{T}$, $\tilde{B}$ and $\tilde{G}$ through translation commutes with the conjugation by $N_{\tilde{G}}(\tilde{T})$, $\tilde{B}$ and $\tilde{G}$. Therefore, if C operates on $\tilde{G} \times^{\tilde{B}} \tilde{B}$, $(c, g * b) \rightarrow g * cb$, and on $\tilde{T}/W$, $(c, \bar{t}) \mapsto \overline{ct}$, the morphisms in 4.3 define a diagram

$$\begin{array}{ccc} \tilde{G} \times^{\tilde{B}} \tilde{B} & \xrightarrow{\tilde{\phi}} & G \\ {\scriptstyle\tilde{\theta}}\downarrow & & \downarrow{\scriptstyle\tilde{\chi}} \\ \tilde{T} & \xrightarrow{\tilde{\psi}} & \tilde{T}/W \end{array}$$

which is equivariant with respect to C . By taking all the quotients by C , we get the diagram

$$\begin{array}{ccc} G \times^{B} B & \xrightarrow{\phi} & G \\ {\scriptstyle\theta}\downarrow & & \downarrow{\scriptstyle\chi} \\ T & \xrightarrow{\psi} & T/W \end{array} \quad .$$

The operation of C on $\tilde{T}$ is free. Therefore the fibers of θ are isomorphic to the fibers of $\tilde{\theta}$. However, in $\tilde{T}/W$ there are points $\bar{t}$ with nontrivial isotropy group $C_{\bar{t}} \subset C$. Let $t \in \tilde{\psi}^{-1}(\bar{t})$, let s be the image of t in T , and $\bar{s}$ the image of s in T/W . Under the projection $\tilde{G} \rightarrow G$, $Z(t) = Z_{\tilde{G}}(t)$ is mapped to

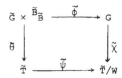

$Z(s)^\circ = Z_G(s)^\circ$ and $\tilde{\chi}^{-1}(\bar{t}) = \tilde{G} \times^{Z(t)} V(t)$ is isomorphic to $G \times^{Z(s)^\circ} V(s)$. Since the group $C_{\bar{t}}$ operates freely on the fiber $\tilde{\chi}^{-1}(\bar{t})$, it follows that $\chi^{-1}(\bar{s}) \cong \tilde{\chi}^{-1}(\bar{t})/C_{\bar{t}}$ and $C_{\bar{t}} \cong Z(s)/Z(s)^\circ$.

The isotropy groups of W on $\tilde{T}$ are reflection groups ([S-S] II.4.2). Comparing now the isotropy groups of C on $\tilde{T}/W$, of W on $T = \tilde{T}/C$, and of $W \times C$ on $\tilde{T}$ we see that $C_{\bar{t}}$ is isomorphic with the group $Z_W(s)/Z_W(s)^\circ$, where $Z_W(s)^\circ = Z_W(s) \cap Z_G(s)^\circ$ (cf. 3.4 and [S-S] II.4.4, 4.7). The group $C_{\bar{t}} \cong Z_W(s)/Z_W(s)^\circ$ is responsible for singularities in T/W.

<u>Example</u>: Let $G = PGL_3$, $\tilde{G} = SL_3$. For $\mathrm{char}(k) \neq 3$, the center C of SL_3 is isomorphic to the group μ_3 of the third roots of unity, which are embedded diagonally in the maximal torus $\tilde{T} \cong \{(x_1,x_2,x_3) \in (k^*)^3 | x_1x_2x_3 = 1\}$ of SL_3. The Weyl group W which is isomorphic to the symmetric group σ_3 operates on $\tilde{T}$ by permutation of the coordinates. The action of C on $\tilde{T}/W$ induced by translation possesses a unique fixed point, namely the W-class of the element $(1,\xi,\xi^2)$ with $\xi \in \mu_3$ (this class contains 6 elements). The group SL_3 is of type A_2. The two fundamental representations of SL_3 are the natural one $\rho_1 : SL_3 \to GL_3$ and the contragradient $\rho_2 = {}^t\rho_1^{-1}$. For the corresponding fundamental characters χ_1 and χ_2 we have $\chi_1(cg) = \xi\chi_1(g)$, and $\chi_2(cg) = \xi^2\chi_2(g)$ for all $g \in SL_3$ and $c = (\xi,\xi,\xi) \in C \cong \mu_3$. Therefore, the quotient $T/W = C\backslash\tilde{T}/W$ can be identified with the quotient of $\mathbb{A}^2$ by the μ_3-action $\mu_3 \times \mathbb{A}^2 \to \mathbb{A}^2$, $(\xi,(\chi_1,\chi_2)) \to (\xi\chi_1,\xi^2\chi_2)$, and so it can be identified using the μ_3-invariants $U = \chi_1 \cdot \chi_2$, $V = \chi_1^3$, $W = \chi_2^3$ with the surface $U^3 = VW$ in $\mathbb{A}^3$. In $\tilde{T}/W \cong \mathbb{A}^2$ the images of $e = (1,1,1)$ resp. the W-orbit of $(1,\xi,\xi^2)$ have the coordinates $\chi_1 = \chi_2 = 3$ resp. $\chi_1 = \chi_2 = 0$, and the corresponding images in $T/W \subset \mathbb{A}^3$ are the points $(9,27,27)$ resp. $(0,0,0)$. The quotient T/W is therefore smooth in a neighborhood of $\bar{e} = (9,27,27)$. On the other hand, the image $(0,0,0)$ of the fixed point in $\tilde{T}/W$ of C is an isolated singularity.

<u>Remark</u>: In general it can be shown that the group C operates freely in an open

neighborhood of $\bar{e}$ in $\tilde{T}/W$ (i.e. the isotropy group $C_{\bar{e}}$ of $\bar{e}$ is trivial), so that the quotient $T/W = C\backslash\tilde{T}/W$ is smooth in a neighborhood of $\bar{e} \in T/W$. Another argument for this fact is that the isotropy group of $e \in T$ under W, namely W, operates locally as a reflection group (cf. also 3.15).

<u>4.6 A Generalization.</u> (This and the next section 4.7 will be used only in 6.6 which again is only needed in 8.10; cf. also the remark 2 in 8.10.) For the simultaneous resolution of the adjoint quotient $\chi : G \to T/W$ we only required the construction of 4.3 in the special case $P = B$. Nevertheless, for $P \neq B$, we also obtain something similar to a simultaneous resolution for certain subvarieties of G, which are known in the theory of enveloping algebras as closures of "Dixmier sheets" ([B-Kr]).

Let G be a reductive group with maximal torus T, Borel subgroup $B \supset T$ and parabolic subgroup $P \supset B \supset T$. The constructions of 4.3 carry over to the reductive case, and we keep the notations $\theta,\chi,\phi,\psi,A,U,C,\ldots$.

Let $D_P = \{gag^{-1} | a \in A,\ g \in G\}$ be the image of $\phi : G \times {}^P A \to G$. Since ϕ is proper, D_P will be closed in G. Let $\chi_P : D_P \to T/W$ be the restriction of χ to D_P. We then have the commutative diagram

$$\begin{array}{ccc} G \times {}^P A & \xrightarrow{\ \phi\ } & D^P \\ \theta \big\downarrow & & \big\downarrow \chi_P \\ C & \xrightarrow{\ \psi\ } & T/W \end{array}$$

Because of 4.3 Lemma 3, the study of this diagram reduces in essence to the study of the morphism $\pi = \phi_e : G \times {}^P U \to D_P \subset V(G)$, after replacing G by reductive subgroups $Z(t) = Z_G(t)$, $t \in C$.

<u>Theorem 1</u> (Richardson, [Ri2], [St2] 3.9 Th. 2, Cor. 2): <u>There is exactly one conjugacy class under</u> P <u>which is open and dense in</u> U. <u>Every element</u> x <u>of this class</u>

is contained in finitely many G-conjugates of U . *The centralizer* $Z_G(x)$ *operates transitively on the set of these conjugates, the identity component* $Z_G(x)^{\circ}$ *lies in* P , *and* $[Z_G(x) : Z_P(x)]$ *is equal to the number of conjugates of* U *which contain* x .

Proof: To the proofs in loc. cit. we only have to add the determination of the number of conjugates of U . Therefore let x be an element of the open (and therefore dense) class in U , and let x be contained in $gUg^{-1} = {}^{g}U$. We may assume $g \in Z_G(x)$. Since $N_G(U) = P$ ([Bo-Ti] 4.) we have ${}^{g}U = U$ exactly when g lies in P . Hence $\# \{{}^{g}U \mid g \in Z_G(x)\} = [Z_G(x) : Z_P(x)]$.

Corollary 1: *There is exactly one dense, open G-orbit in* $G \times {}^{P}U$. *Its image under* π *is a dense, open G-orbit of* $D_P \cap V(G) = \chi_P^{-1}(\bar{e})$. *The restriction of* π *to the dense orbit of* $G \times {}^{P}U$ *induces a cover over the image of separable degree* $[Z_G(x) : Z_P(x)]$ *where* x *lies in the dense P-orbit of* U . *In particular,* $G \times {}^{P}U$ *and* $\chi_P^{-1}(\bar{e})$ *are irreducible of dimension* $\dim G - \dim P + \dim U = 2 \dim U$.

Proof: The first statement is clearly true. The morphism $\pi : G \times {}^{P}U \to D_P \cap V(G)$ factors over the embedding τ (3.7)

$$\begin{array}{ccc} G \times {}^{P}U & \xrightarrow{\quad \pi \quad} & V(G) \\ & {\scriptstyle \tau} \searrow \quad \nearrow {\scriptstyle p_2} & \\ & G/P \times V(G) & \end{array}$$

.

Because $N_G(U) = P$, we can identify G/P as the variety $\underline{U}$ of all G-conjugates of U . The fiber $\tau(\pi^{-1}(x))$ of p_2 over $x \in U \subset V(G)$ is then equal to $\{(U',x) \in \underline{U} \times \{x\} \mid x \in U'\}$, which has cardinality $[Z_G(x) : Z_P(x)]$. The image $\pi(G * x)$ is now dense in $\pi(G \times {}^{P}U) = \chi_P^{-1}(\bar{e})$ and is open since it is the complement of finitely many smaller orbits in $D_P \cap V(G)$. The last assertions now follow.

Corollary 2: Let G be semisimple, simply connected and char(k) very good for G. Every fiber $\theta^{-1}(t)$, $t \in C$, of θ contains exactly one dense, open G-orbit, whose image under ϕ_t is open in $\chi_P^{-1}(\bar{t})$. The restriction of ϕ_t to this orbit is ètale over the image with degree $[Z_{Z(t)}(x) : Z_{P(t)}(x)]$, where x is an element in the dense $P(t)$-orbit of $U(t)$. The fibers of θ are irreducible of dimension $2 \dim U$, and those of χ_P are the union of equal dimensional components of dimension $2 \dim U$.

Proof: The statements follow from 4.3 Lemma 3 and Corollary 1 above (cf. also the proof of 4.4). The separability of ϕ_t follows from that of the orbit maps (cf. 3.13).

Remark: The fibers $\chi_P^{-1}(\bar{t})$, $t \in C$, are not necessarily irreducible (e.g. the "subregular" examples in 6.5). In general

$$\chi_P^{-1}(\bar{t}) = \bigcup_{s \in \psi^{-1}(\bar{t})} \phi_s(\theta^{-1}(s)) .$$

For the case $P = B$, the images $\phi_s(\theta^{-1}(s))$, $s \in \psi^{-1}(\bar{t})$, all coincide.

If $\Delta_P \subset \Delta$ is the subsystem of simple roots of T in G defined by P, and Σ_P is the root subsystem of Σ generated by Δ_P, then C is defined by $C = \{t \in T \mid \alpha(t) = 1 \text{ for all } \alpha \in \Delta_P\}$. The elements of the dense and open subset $C^{reg} := \{t \in T \mid \alpha(t) \neq 1 \text{ for all } \alpha \in \Sigma \setminus \Sigma_P\}$ of C are called the regular elements of C. Obviously the identity component $Z_G(t)^\circ$ of the centralizer of a regular element $t \in C$ is just the reductive part M of P (cf. 3.4, 4.3). We let ${}^GC = \{gcg^{-1} \mid g \in G, c \in C\}$.

Lemma 1: The closure in G of ${}^GC$ as well as that of ${}^GC^{reg}$ is equal to D_P.

Proof: Since $U(t) = Z(t) \cap U = Z(t)^\circ \cap U = M \cap U = \{e\}$ for $t \in C^{reg}$, the fiber $\theta^{-1}(t)$ is isomorphic to $G/M = G/Z(t)^\circ$ (4.3 Lemma 3), and $\phi(\theta^{-1}(t))$ is just the G-orbit of t in G. Due to the fact that θ is smooth (and hence open) $\theta^{-1}(C^{reg})$

is open and dense in $G \times {}^{P}A$. Therefore ${}^{G}C^{reg} = \phi(\theta^{-1}(C^{reg}))$ is also dense in $D_P = \phi(G \times {}^{P}A)$.

Now let P' be another parabolic subgroup of G which contains T and B . Let $\Delta_{P'} \subset \Delta$ be the subsystem determined by P' . Let $C' = \{t \in T \mid \alpha(t) = 1$ for all $\alpha \in \Delta_{P'}\}$.

__Corollary 3:__ __We have__ $D_P = D_{P'}$ __exactly when__ Δ_P __and__ $\Delta_{P'}$ __are conjugate as subsets of__ Σ __under the Weyl group__ W .

__Proof:__ If Δ_P and $\Delta_{P'}$ are conjugate under W , so are C and C' . By the Lemma, we will have $D_P = D_{P'}$. On the other hand, suppose $D_P = D_{P'}$. By the Lemma, ${}^{G}C^{reg} = {}^{G}C'^{reg}$. Then there is a regular element $t \in C$ which is conjugate under G to a regular element $s \in C'$. Since s and t both lie in T , they will be conjugate under W , as will their centralizers $Z_G(t)$ and $Z_G(s)$, and the corresponding centers C and C' . Therefore Δ_P and $\Delta_{P'}$ are also conjugate under W .

Now let P_i , $i \in \{1,\dots,r\}$ be the minimal parabolic subgroups $\neq B$ of G defined by $\Delta_{P_i} = \{\alpha_i\} \subset \Delta$, $i \in \{1,\dots,r\}$ (r = rank G) . Let $D_i = D_{P_i}$.

__Theorem 2__ ([St1] 5.1): __An element__ $x \in G$ __is irregular__ __(i.e. not regular)__ __exactly when__ x __lies in the union__ $\bigcup_{i=1}^{r} D_i$.

__Corollary 4:__ __If__ G __is simple,__ __then__ $D_i = D_j$ __exactly when__ α_i __and__ α_j __are roots of equal length.__

__Proof:__ It is easy to verify that roots of equal length in an irreducible root system are conjugate under W . The assertion then follows from Corollary 3 .

__Remarks:__ 1) If G is simple with homogeneous Dynkin diagram, then all the D_i coincide. If the diagram is inhomogeneous, then $\bigcup_{i=1}^{r} D_i = D_j \cup D_k$ where α_j is a short and α_k a long root.

2) The orbits of maximal dimension in $\bigcup_{i=1}^{r} D_i$ have, by Corollary 2, dimension $\dim G - r - 2$, and are therefore subregular in the sense of 5.4.

<u>4.7. Carrying over to Lie Algebras.</u> Let G be semisimple with Lie algebra $\underline{g}$ and char(k) very good. There are statements for $\underline{g}$ analogous to the results of 4.1, 4.4 and 4.6. Let $\underline{b}$ resp. $\underline{h}$ be the Lie algebras of B resp. T . From 4.4 we get an analogously defined commutative diagram

$$\begin{array}{ccc} G \times {}^{B}\underline{b} & \longrightarrow & \underline{g} \\ \big\downarrow & & \big\downarrow \gamma \\ \underline{h} & \longrightarrow & \underline{h}/W \end{array}$$

which is a simultaneous resolution of γ . The proof proceeds as for the results in 3.14, only by means of the additive Jordan decomposition in $\underline{g}$ instead of the multiplicative one in G . In the same way the statements of 4.6 can be transcribed. (For $\mathrm{char}(k) = 0$ there is a somewhat different derivation in [B-Kr].) The results cited from Richardson are already formulated for $\underline{g}$ in [Ri], and the characterization of the irregular elements (Theorem 2) for $\underline{g}$ can be found in [Ve] 4.7. Besides the analogy between G and $\underline{g}$, there is a stronger connection between the resolution

$$\pi : G \times {}^{B}U \longrightarrow V(G)$$

of the unipotent variety and the resolution

$$\pi' : G \times {}^{B}\underline{n} \longrightarrow N(\underline{g})$$

of the nilpotent variety (here $\underline{n} = \mathrm{Lie}\ U$). According to [Sp2] (proof of 3.1, and

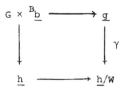

[Ve] 6.9) there is a B-equivariant isomorphism $U \to \underline{n}$ which induces compatible G-isomorphisms α and β

$$\begin{array}{ccc}
G \times {}^{B}U & \xrightarrow{\ \pi\ } & V(G) \\
{\scriptstyle\alpha}\big\downarrow & & \big\downarrow{\scriptstyle\beta} \\
G \times {}^{B}\underline{n} & \xrightarrow{\ \pi'\ } & N(\underline{g})
\end{array}$$

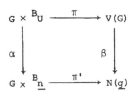

III Simple Singularities in Simple Groups

In this part, k will be an algebraically closed commutative field.

5. Subregular Singularities

5.1. Transverse Slices. Let $V(G)$ be the unipotent variety of the semisimple group G. We will study the singularities of $V(G)$ along the unipotent orbits. Along an orbit, $V(G)$ possesses a certain homogeneity and the essential geometric properties will be reflected in a transverse slice to the orbit.

Definition: Suppose the algebraic group G operates regularly on the variety X. A transverse slice in X to the orbit of $x \in X$ at the point x will be a locally closed subvariety S of X with the properties

i) $x \in S$,

ii) the morphism $G \times S \to X$, $(g,s) \mapsto g \cdot s$, is smooth,

iii) the dimension of S is minimal with respect to satisfying (i) and (ii).

Lemma 1: If X is a smooth variety with a regular G-action, then for every $x \in X$ there is a transverse slice S to the orbit of x in X. Moreover let H be a linearly reductive subgroup of the stabilizer $Z_G(x) = \{g \in G \mid gx = x\}$ and assume that either X is affine or H is finite. Then S may be chosen to be stable with respect to the natural action of H on X.

Proof (Compare [Lu] III.1 Lemme): Because of our assumptions there is an H-stable affine open neighborhood U of x in X. Let $\underline{m} \subset k[U]$ denote the H-stable maximal ideal defining x in U. Since H is linearly reductive the H-equivariant natural projection $\underline{m} \to \underline{m}/\underline{m}^2$ admits an H-equivariant section s, which induces an H-equivariant homomorphism $S(s) : S_*(\underline{m}/\underline{m}^2) \to k[U]$. Geometrically $S(s)$ corresponds to an H-equivariant morphism $\pi : U \to T_xX$ from U to the tangent space of X in x, which is étale at x. Let $E \subset T_xX$ be the image of the Lie algebra of G under the differential at e of the orbit map $G \to X$, $g \mapsto gx$. Then E is stable under H. Now let $S \subset U$ be the preimage under π of some linear H-complement to E in $T_x X$. Then $s \in S$, and S is smooth at x. By construction, $G \times S \to X$, $(g,s) \mapsto gs$, is smooth at $(e,x) \in G \times S$ (EGA IV 17.11.1). Since this morphism is equivariant with respect to G (G operates on $G \times S$ by left translations on the left factor) and is invariant with respect to the H-action $H \times G \times S \to G \times S$, $(h,g,s) \mapsto (gh^{-1},hs)$, we can, after replacing S by an H-stable open subvariety of S, assume that $G \times S \to X$ is smooth at all points. Since S has the minimal dimension that will satisfy these properties, S will be the desired transverse slice.

<u>Remarks</u>: 1) The existence of transverse slices in singular varieties also follows from Lemma 2 (see below) in conjunction with 1.3.

2) If X is smooth, every transverse slice S in X must be smooth (EGA IV 17.5.8). The dimension of S is greater or equal to the codimension of the orbit $G \cdot x$ in X, equality holding exactly when the orbit map from G to $G \cdot x$ is separable. This will occur, for example, for $\mathrm{char}(k) = 0$ or for the adjoint action of a semisimple group in very good characteristic (e.g. 3.13).

3) Assume that the orbit map $G \to G \cdot x$ is separable and that $Z_G(x)$ is linearly reductive. If S is a $Z_G(x)$-stable transverse slice to the orbit of x at the point x, then the morphism $G \times S \to X$ will factor through a morphism $G \times^{Z_G(x)} S \to X$ which will be étale for dimension reasons. If X is affine and $G \cdot x$ closed, then this situation provides an analogon to the "normal slice theorem" for compact Lie group actions (cf. [Jä], [Lu]). Later on we have to deal with transverse slices to the adjoint orbit of a unipotent element x in a reductive group G. In general $Z_G(x)$ will not be linearly reductive (cf. 7.5), and there will be no $Z_G(x)$-stable transverse slice to the orbit of x (for an example of the difficulties, consider the usual action of the upper triangular matrices in GL_n on $\mathbb{A}^n$).

Lemma 2: *Suppose the algebraic group* G *operates regularly on the varieties* X *and* Y. *Let* $\phi : X \to Y$ *be a G-equivariant morphism. In addition, let* S *be a (locally closed) subvariety of* X *and* $S' = S \times_X Y$ *be the preimage of* S *under* ϕ. *Then the following diagram is cartesian*

$$\begin{array}{ccc} G \times S' & \xrightarrow{\mu'} & Y \\ {\scriptstyle id_G \times \phi}\downarrow & & \downarrow{\scriptstyle \phi} \\ G \times S & \xrightarrow{\mu} & X \end{array}$$

where $\mu(g,s) = gs$ *and* $\mu'(g,s') = gs'$. *If* μ *is smooth, so is* μ'.

Proof: In the following commutative diagram the vertical arrows are induced by $\phi : Y \to X$ and $id : G \to G$, and λ resp. λ' are defined by $(g,u) \mapsto (g,gu)$ with $u \in X$ resp. $u \in Y$.

$$\begin{array}{ccccccc} G \times S' & \longrightarrow & G \times Y & \xrightarrow[\lambda']{\sim} & G \times Y & \xrightarrow[p_2]{} & Y \\ \downarrow & & \downarrow & & \downarrow & & \downarrow \\ G \times S & \longrightarrow & G \times X & \xrightarrow[\lambda]{\sim} & G \times X & \xrightarrow[p_2]{} & X \end{array}$$

The composition of the top resp. bottom line gives μ resp. μ'. The first statement follows since each square is cartesian. Consequently the morphism μ' is induced from μ by the base change ϕ. Therefore if μ is smooth, so is μ' (EGA IV 17.3.3).

Example: If $\phi : Y \to \mathbb{A}^n$ is an embedding of a possibly singular G-variety into a linear G-action and S is a transverse slice in $\mathbb{A}^n$ to the orbit of $\phi(y)$, where $y \in Y$, then $\phi^{-1}(S)$ is a transverse slice in Y to the orbit of y.

We now wish to compare different slices to an orbit.

Definition: Let $S_1 \to Y$ and $S_2 \to Y$ be varieties over a variety Y, and let

$x_1 \in S_1$, and $x_2 \in S_2$ be points over $y \in Y$. Then the pairs (S_1,x_1) and (S_2,x_2) are called locally isomorphic over Y in the ètale topology if there are a variety S over Y with a point s over y and ètale Y-morphisms $\phi_i : S \to S_i$ with $\phi_i(s) = x_i$, $i = 1,2$. If (S_1,x_1) and (S_2,x_2) are locally isomorphic over Y in the ètale topology then the Henselizations (and the completions) of S_1 at x_1 resp. of S_2 at x_2 are isomorphic over Y (cf. EGA IV 18.6). In case $k = \mathbb{C}$ "locally isomorphic in the ètale topology" implies "locally analytic isomorphic" (for the usual Hausdorff topology).

Lemma 3: Suppose G operates regularly on the variety X and trivially on the variety Y . Let $\phi : X \to Y$ be a G-equivariant morphism, and let S_1 and S_2 be transverse slices to an orbit of G in X at points $x_1 \in S_1$ and $x_2 \in S_2$ respectively. Then (S_1,x_1) and (S_2,x_2) are locally isomorphic over Y in the ètale topology. Let moreover H be a linearly reductive subgroup of $Z_G(x_1)$, and assume that S_1 is stable under H and that S_2 is stable under gHg^{-1} for some $g \in G$ with $gx_1 = x_2$. Then the isomorphism above can be chosen equivariant with respect to the canonical isomorphism $H \cong gHg^{-1}$.

Proof: After applying g^{-1} to (S_2,x_2) we may assume $x_1 = x_2 = x$ and S_2 stable under H . We let H act on the Lie algebra T_eG of G by the restriction to H of the adjoint representation. Let C be an H-stable linear complement to the infinitesimal centralizer of x , i.e. to the Lie subalgebra of the (not necessarily reduced) centralizer scheme $Z_G(x) = \{g \in G \mid gx = x\}$. Choose an H-equivariant projection $\pi : G \to T_eG$ which is ètale at e (cf. Proof of Lemma 1) and consider $C_1 = \pi^{-1}(C)$, $C_2 = C_1^{-1} = \{c^{-1} \mid c \in C_1\}$. Then C_1 and C_2 are stable with respect to conjugation by H . Define an H-action on $C_i \times S_i$, $i = 1,2$, by $(h,c,s) \mapsto (hch^{-1},hs)$. Then the morphisms

$$m_i : C_i \times S_i \longrightarrow X , \; m_i(c,s) = cs, \; i = 1,2 ,$$

will be H-equivariant and ètale in H-stable neighborhoods U_i of (e,x) in $C_i \times S_i$.

(cf. EGA IV 17.16. 1-3). Let $M_1 = m_2^{-1}(S_1)$ and $M_2 = m_1^{-1}(S_2)$. Then the H-equivariant projections $M_1 \cap U_2 \to S_1$ and $M_2 \cap U_1 \to S_2$ are ètale (base change). Since

$$M_1 = (C_2 \times S_2) \times_X S_1 \longrightarrow (C_1 \times S_1) \times_X S_2 = M_2$$

$$(c,s,t) \longmapsto (c^{-1},t,s)$$

is an H-equivariant isomorphism over Y, we are done.

5.2. The Restriction of Invariants to Transverse Slices.

Again consider the adjoint action of a semisimple group G on itself. Without harming later investigations we can choose G to be simply connected. Then the adjoint quotient $\chi : G \to T/W$ is flat. Let $x \in G$, let $S \subset G$ be a transverse slice to the orbit of x at the point x and let $\sigma : S \to T/W$ be the restriction of χ to S. Then:

Lemma: The morphism σ is flat, its fibers are normal, and $x \in S$ is a regular point of the fiber $\sigma^{-1}\sigma(x)$ exactly when x is a regular element in G.

Proof: The composition of the smooth (and therefore flat) morphism $\mu : G \times S \to G$, $(g,s) \mapsto gs$, with the flat morphism χ is again flat. Because of its G-invariance, the composition factors through the projection to S

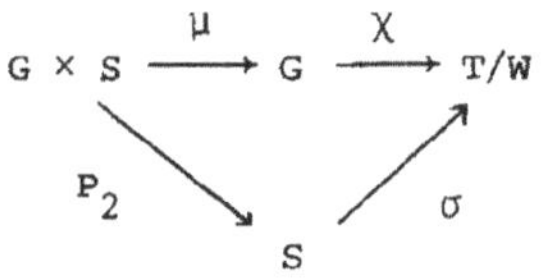

The projection P_2 is surjective and flat. Thus the flatness of σ follows (for example, use EGA IV 2.2.11). By 5.1 Lemma 2 the morphism $\mu_{\bar{t}} : G \times \sigma^{-1}(\bar{t}) \to \chi^{-1}(\bar{t})$ is smooth for all $\bar{t} \in \sigma(S) \subset T/W$, and a point $x \in \sigma^{-1}(\bar{t})$ is normal resp. regular exactly when the point (e,x) in $G \times \sigma^{-1}(\bar{t})$ is, and by EGA IV 17.5.8, exactly when the point $\mu_{\bar{t}}(e,x) = x$ in $\chi^{-1}(\bar{t})$ is. The rest follows by Theorem 3.10.

Remark 1: An analogous statement obviously holds when we consider the adjoint representation of G on the Lie algebra $\underline{g}$ for very good characteristic (cf. 3.14).

Remark 2: As the image of the flat morphism σ, $\sigma(S) \subset T/W$ is an open set (EGA IV 2.4.6).

5.3. Simultaneous Resolutions for Transverse Slices.

We consider the situation of 5.2 and will show how the simultaneous resolution of χ in 4.4

$$\begin{array}{ccc} G \times^{B} B & \xrightarrow{\phi} & G \\ \theta \downarrow & & \downarrow \chi \\ T & \xrightarrow{\psi} & T/W \end{array}$$

leads to a simultaneous resolution of σ. Let $S' = (G \times^{B} B) \times_G S$ be the preimage $\phi^{-1}(S)$ of the transverse slice S and $\theta' : S' \to T$ resp. $\phi' : S' \to G$ the corresponding restrictions of θ and ϕ on S'.

Corollary (to 4.4): The following diagram is a simultaneous resolution of σ

$$\begin{array}{ccc} S' & \xrightarrow{\phi'} & S \\ \theta' \downarrow & & \downarrow \sigma \\ T & \xrightarrow{\psi} & T/W \end{array}$$

Proof: Since ϕ is proper, so is ϕ' (base change, EGA II 5.4.2). We get the smoothness of θ' as follows. Applying 5.1 Lemma 2 to $\phi : G \times^{B} B \to G$, it follows that $\mu' : G \times S' \to G \times^{B} B$, $(g,s') \mapsto gs'$ is smooth. Since $\theta : G \times^{B} B \to T$ is

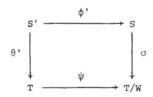

G-invariant, the smooth composition $\theta \circ \mu'$ factors through the smooth projection p_2 of $G \times S'$ to S'. Then by EGA IV 17.11.1, θ' is also smooth:

$$\begin{array}{ccccc} G \times S' & \xrightarrow{\mu'} & G \times^B B & \xrightarrow{\theta} & T \\ & {\scriptstyle p_2} \searrow & & \nearrow {\scriptstyle \theta'} & \\ & & S' & & \end{array}$$

In particular, the fibers of θ' and S' itself are smooth (EGA IV 17.5.8). Now it only remains to show that ϕ' induces a resolution $\phi'_t : S'_t = \theta'^{-1}(t) \to S_{\bar{t}} = \sigma^{-1}(\bar{t})$ for all $t \in \psi^{-1}(\sigma(S)) \subset T$ and $\bar{t} = \psi(t)$. By 5.1 Lemma 2 the following diagram is cartesian

$$\begin{array}{ccc} G \times S'_t & \xrightarrow{\mu'_t} & \theta^{-1}(t) \\ {\scriptstyle id_G \times \phi'_t} \downarrow & & \downarrow {\scriptstyle \phi_t} \\ G \times S_{\bar{t}} & \xrightarrow{\mu_t} & G_{\bar{t}} \end{array}$$

Because G is smooth, ϕ'_t will be a resolution exactly when $id_G \times \phi'_t$ is a resolution. This last follows in a simple way from the smoothness of $\mu_{\bar{t}}$ which just implies that $\mu_{\bar{t}}$ maps the regular resp. singular points of $G \times S_{\bar{t}}$ to the regular resp. singular points of $G_{\bar{t}}$ (EGA IV 17.5.8). Therefore $id_G \times \phi'_t$ is an isomorphism over the regular points $(G \times S_{\bar{t}})^{reg}$ (base change). The preimage $(id_G \times \phi'_t)^{-1}(G \times S_{\bar{t}})^{reg}$ is open and dense in $G \times S'_t$, since it is equal to the preimage under the smooth, hence open, morphism μ'_t of the set $\phi_t^{-1}(G_{\bar{t}}^{reg})$ which is open und dense in $\theta^{-1}(t)$. The properness of ϕ_t implies that of $id_G \times \phi'_t$ (base change). With that the corollary is proven.

<u>Remark:</u> The morphism $\phi' : S' \to S$ of the simultaneous resolution defined above induces in particular a resolution $\pi' = \phi'_e : S'_e \longrightarrow S_{\bar{e}}$ of the fiber $S_{\bar{e}} = \sigma^{-1}(\bar{e})$. in 4.1 we have seen that the reduced exceptional set $E(\pi)$ of $\pi = \phi_e$,

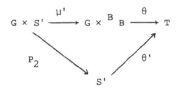

$\phi_e : G \times {}^B U \to V(G)$ is a divisor with normal crossings. Since the slice S' intersects any G-orbit it meets transversely, it intersects all G-stable submanifolds of $G \times {}^B U$ transversely too. Hence the intersections of S' with the components of $E(\pi)$ as well as with the iterated intersections of these components are again manifolds in S', i.e. the reduced exceptional set $E(\pi') = E(\pi) \cap S'$ of π' is likewise a divisor with normal crossings. This can also be derived formally: By 5.1 Lemma 2 we get a cartesian diagram

$$\begin{array}{ccc} G \times E(\pi') & \longrightarrow & E(\pi) \\ \downarrow & & \downarrow \\ G \times S'_e & \longrightarrow & G \times {}^B U \end{array}$$

in which the horizontal arrows will be induced by μ'_e. Because μ'_e is smooth, $G \times E(\pi')$ in $G \times S'_e$, and hence $E(\pi')$ in S'_e will both be divisors with normal crossings.

5.4. Subregular Elements.

In the following, we will only consider those special orbits of a semisimple group which are related to the simple singularities. In general, let G be a reductive group of rank r. The minimal centralizer dimension of an element $x \in G$ is then r (cf. 3.8) and is obtained exactly for the regular elements of G. Theorem 3.10 implies that the next larger centralizer dimension must be equal to or bigger than $r + 2$ (the statement in 3.10 is formulated for semisimple groups, but extends immediately to the reductive case). A general result says that all orbits in G will have even dimension when the characteristic of k is good for G (cf. [S-S] III 3.25).

Definition: An element $x \in G$ is called subregular exactly when the centralizer $Z_G(x)$ has dimension $r + 2$.

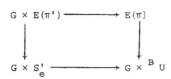

Theorem ([St2] 3.10 Th. 1): In a simple algebraic group G there is exactly one conjugacy class of subregular unipotent elements. This class is dense in the complement of the regular class in the unipotent variety of G.

Again let G be reductive and let $x = x_s \cdot x_u$ be the Jordan decomposition of an element $x \in G$. By 3.3 and 3.4 we have $Z_G(x) = Z_{Z_G(x_s)}(x_u)$ and $x_u \in Z_G(x_s)^{\circ}$. Furthermore, $Z_G(x_s)^{\circ}$ is reductive of the same rank as G. Therefore we have

Lemma 1: An element $x \in G$ is subregular exactly when its unipotent part x_u is a subregular element of the reductive group $Z_G(x_s)^{\circ}$.

Since the unipotent elements of a reductive group lie in its semisimple commutator, it suffices to consider subregular unipotent elements in semisimple groups. Such a group is the almost direct product of its simple normal subgroups $G = G_1 \cdot \ldots \cdot G_m$, i.e. the homomorphism $\mu : G_1 \times \ldots \times G_m \to G$, $\mu(g_1, \ldots, g_m) = g_1 \cdot \ldots \cdot g_m$ is surjective, separable, has a finite central kernel, induces an isomorphism of the unipotent varieties and respects the centralizerdimensions (cf. [Bo] IV 14.10). The following statement is now obvious.

Lemma 2: Let $u = u_1 \cdot \ldots \cdot u_m$ be a unipotent element of G with $u_i \in G_i$, $i=1,\ldots,m$. Then u is subregular exactly when there is a $j \in \{1,\ldots,m\}$ such that u_j is subregular in G_j, and the u_i, $i \neq j$, are regular in G_i. For every simple normal subgroup of G there is exactly one corresponding subregular unipotent orbit in G.

Definition: Let G be as above, $u = u_1 \cdot \ldots \cdot u_m$ be a subregular unipotent element, where $u_i \in G_i$ is subregular. If Δ_i is the Dynkin diagram of G_i, then u is called subregular of type Δ_i.

Remark: Analogous definitions and statements can be made for the Lie algebra of G when the characteristic of k is very good (cf. 3.13, 3.14).

5.5. Subregular Singularities.

Let G be a simple and simply connected group. If the characteristic of k is good for G, then the orbit maps of the adjoint group for the adjoint action are separable. (We let the adjoint group of G act rather than G itself to get rid of the inseparability in the case when $G = GL_n$, $\mathrm{char}(k) | n$.) The separability of the orbit maps implies the geometrically correct dimension for the transverse slices, i.e. the dimension of the slices is equal to the codimension of the concerned orbits.

Let $\mathrm{char}(k)$ be good and S be a transverse slice to the subregular orbit of G, which now has dimension $r + 2$ where r is the rank of G. Since S is transverse to the G-orbits at all its points, it can intersect only regular or subregular orbits. Now the fibers of the morphism $\sigma : S \to T/W$ are normal surfaces (5.2), whose isolated singularities correspond to subregular elements in G. We will consider the desingularization of these surfaces given by the simultaneous resolution in 5.3. As was remarked there, the reduced exceptional set is a divisor with normal crossings. Therefore it is a system of finitely many smooth and complete curves, which intersect transversely. By a result of Tits ([Sp2] 1.5, [St2] 3.9 Prop. 1, see also 6.2) two points of the exceptional set E can be connected in E by a connected system of rational curves $(\cong \mathbb{P}^1)$. Therefore the components of the exceptional divisor itself must be isomorphic to $\mathbb{P}^1$. Because of more results on this resolution in the following sections we will be able to identify the corresponding singularities.

6. Simple Singularities

6.1. Rational Double Points. In this chapter we will prove that the subregular singularities of section 5.5 are a class of singularities known as rational double points in algebraic and analytic geometry where they have been studied in detail.

Let (X,x) be the spectrum of a two-dimensional normal local k-algebra. Here we let $x \in X$ denote the closed point of X. Then (X,x) is called rational when for a resolution $\pi : X' \longrightarrow X$ of the singularities of X, the higher direct images vanish, $R^q\pi_*(\mathcal{O}_{X'}) = 0$ for $q \geq 1$ (cf. [Ar2]). It can be shown that this definition is independent of the choice of a resolution of π (loc. cit.).

If $\pi : X' \to X$ is a resolution, then the reduced exceptional divisor $E = \pi^{-1}(x)_{red}$ consists of a system of irreducible curves $E = C_1 \cup \ldots \cup C_r$. We wish to distinguish a certain class of the possible configurations of these curves.

Definition: A Dynkin diagram is homogeneous, when the corresponding root system contains only roots of the same length.

The Cartan matrix for a homogeneous Dynkin diagram Δ, $((n_{\alpha,\beta}))$, $\alpha,\beta \in \Delta$ (we identify Δ with the set of vertices without fear of confusion), satisfies $n_{\alpha,\alpha} = 2$ for all $\alpha \in \Delta$ and $n_{\alpha,\beta}$ $\{0,-1\}$ for all $\alpha \neq \beta \in \Delta$. The irreducible homogeneous Dynkin diagrams are just A_r, $r \geq 1$, D_r, $r \geq 4$, E_6, E_7, E_8 (cf. 3.1 and LIE VI).

Let Δ be one such irreducible homogeneous diagram, and $\pi : X' \to X$ a resolution of the surface X (as above).

Definition: The resolution π has an exceptional configuration of type Δ, when the following conditions hold:

i) there is a bijection $\alpha \mapsto C_\alpha$ from the vertices $\alpha \in \Delta$ to the components of the exceptional divisor E of π,

ii) all the C_α, $\alpha \in \Delta$, are isomorphic to the projective line $\mathbb{P}^1$,

iii) the intersection numbers $C_\alpha \cdot C_\beta$ of the components of E are given by the negative Cartan matrix, $C_\alpha \cdot C_\beta = -n_{\alpha,\beta}$ for all $\alpha,\beta \in \Delta$.

Comment: Property (iii) says that the different (smooth) components of E intersect, if at all, transversely at a single point and that the normal bundle of a component of E in X' (after the identification of E with $\mathbb{P}^1$) is isomorphic to the cotangent bundle $T^*\mathbb{P}^1$ or, equivalently, the line bundle $\mathcal{O}(-2)$ and therefore has first Chern class -2 (on the basis of the usual identification of the component $A^1(\mathbb{P}^1)$ of the Chow ring $A(\mathbb{P}^1)$ with $\mathbb{Z}$, see for example [G] 4-12).

We will symbolize a configuration of type Δ by its dual diagram, in which the vertices of Δ are lines and the edges of Δ correspond to transverse intersections, e.g.:

Remark: For a surface X there is a minimal resolution, unique up to isomorphism, through which all other resolutions must factor. The minimal resolution is characterized by the fact that it has no exceptional curves of the first kind, i.e. rational curves with self-intersection -1 (cf. [Br1] Lemma 1.6, [Li] 27.3). Therefore, if the resolution $\pi : X' \to X$ of the surface X has an exceptional configuration of type Δ, then it is minimal.

Theorem ([Ar2], [Br1], [Li] 23.5): The following properties of a normal surface (X,x) are equivalent:

i) (X,x) is rational of embedding dimension 3 at x.

ii) (X,x) is rational of multiplicity two at x.

iii) (X,x) is of multiplicity two at x and it can be resolved by succesive blowing up of points.

iv) The minimal resolution of (X,x) has the exceptional configuration of a Dynkin diagram of type A_r, D_r, E_6, E_7 or E_8.

Definition: If any (hence all) of the four properties of this theorem is satisfied, then (X,x) is called a rational double point or a simple singularity. According to the type of the Dynkin diagram associated to its minimal resolution it is called of type A_r, D_r, E_6, E_7 or E_8.

Theorem ([Ar5]): Let the characteristic of k be good for the homogeneous irreducible Dynkin diagram Δ. Then there is exactly one rational double point of type Δ up to isomorphism of Henselizations. Representatives of the individual classes are given by the local varieties at $0 \in \mathbb{A}^3$ defined by the following equations:

A_r	$X^{r+1} + YZ = 0$
D_r $r \geq 4$	$X^{r-1} + XY^2 + Z^2 = 0$
E_6	$X^4 + Y^3 + Z^2 = 0$
E_7	$X^3Y + Y^3 + Z^2 = 0$
E_8	$X^5 + Y^3 + Z^2 = 0$

Over $\mathbb{C}$ the equations above appeared in the last century in the works of H. A. Schwarz and F. Klein (cf. [Kl]). The following description of the rational double points as quotient singularities also goes back to Klein, in fact they are often called Kleinian singularities.

Theorem ([Kl], [Ar5]): Let the characteristic k be good for the diagram Δ. The equations of the previous theorem give the unique relation (syzygy) between three suitable chosen generators X,Y,Z of the algebra $k[\mathbb{A}^2]^F$ of the invariant polynomials of $\mathbb{A}^2$ under the action of a finite subgroup $F \subset SL_2$; where F is as follows depending on the type of Δ :

A_r : $\mathbb{Z}_{r+1}$, cyclic group of order $r+1$

D_r : $\mathbb{D}_{r-2}$, binary dihedral group of order $4(r-2)$

E_6 : $\mathbb{T}$, binary tetrahedral group of order 24

E_7 : $\mathbb{O}$, binary octahedral group of order 48

E_8 : $\mathbb{I}$, binary icosahedral group of order 120 .

If $char(k) = 0$, these groups are (up to conjugation) the unique finite subgroups of SL_2 .

Therefore, in good characteristic every rational double point is, after Henselization at the singular point, isomorphic to the corresponding quotient $\mathbb{A}^2/F$.

In positive characteristic we have to regard the groups $\mathbb{Z}_{r+1}$ and $\mathbb{D}_{r-2}$ as possibly non reduced group schemes (contrarily to our general conventions 1.1). These groups will not be reduced when $p|r+1$ resp. $p|4(r-2)$. To avoid ambiguity and because of later calculations we will give representatives for the classes of the finite groups in question (cf. [Ar5], [Kl], [Sp3]). In the following discussion we assume $char(k)$ to be good with respect to the Dynkin diagram attached to F by the theorem above.

For $\mathbb{Z}_n$ we choose the group scheme consisting of the n-th roots of unity embedded in the diagonal torus of SL_2 :

$$\begin{pmatrix} \xi & 0 \\ 0 & \xi^{-1} \end{pmatrix} , \ \xi \in \mu_n = \text{Spec } k[T]/(T^n-1) .$$

The group $\mathbb{D}_n$ is generated by $\mathbb{Z}_{2n}$ and the element $\begin{pmatrix} 0 & 1 \\ -1 & 0 \end{pmatrix}$ in SL_2 which normalizes $\mathbb{Z}_{2n}$.

The group $\mathbb{T}$ is generated by the group $\mathbb{D}_2$ and $\frac{1}{\sqrt{2}}\begin{pmatrix} \varepsilon^7 & \varepsilon^7 \\ \varepsilon^5 & \varepsilon \end{pmatrix}$, where ε is a primitive 8-th root of unity. We get the exact sequence

$$1 \to \mathbb{D}_2 \to \mathbb{T} \to \mathbb{Z}/(3) \to 1 .$$

If we enlarge $\mathbb{T}$ by the products with the element $\begin{pmatrix} \varepsilon & 0 \\ 0 & \varepsilon^7 \end{pmatrix}$ of SL_2, where ε is the above 8-th root of unity, we obtain the group $\mathbb{O}$ and the following two exact sequences

$$1 \to \mathbb{T} \to \mathbb{O} \to \mathbb{Z}/(2) \to 1$$

$$1 \to \mathbb{D}_2 \to \mathbb{O} \to \mathfrak{S}_3 \to 1$$

where $\mathfrak{S}_3$ is the symmetric group on 3 letters.

Finally the icosahedral group $\mathbb{I}$ is generated by the matrices $-\begin{pmatrix} \eta^3 & 0 \\ 0 & \eta^2 \end{pmatrix}$ and $\frac{1}{\eta^2-\eta^3}\begin{pmatrix} \eta+\eta^4 & 1 \\ 1 & -\eta-\eta^4 \end{pmatrix}$ where η is a primitive 5-th root of unity.

When $k = \mathbb{C}$ the above groups, because of their finiteness, lie in the compact subgroup $SU_2(\mathbb{C})$ of $SL_2(\mathbb{C})$ which projects as a double cover onto the usual Euclidean orthogonal group $SO_3(\mathbb{R})$. Under this projection the binary dihedral, tetrahedral, octahedral, and icosahedral groups map onto the symmetry groups of the corresponding regular solids (cf. [Kl], [Sp3]).

Remarks: 1) In bad characteristic the uniqueness of the rational double points gets lost. In addition their description as quotient singularities fails. Nevertheless, there exist smooth covers of the singularities (cf. [Ar5], [Li] 25.2).

2) Over $\mathbb{R}$ the functions corresponding to the equations of type $A_1,\dots,A_5$ and D_4, D_5 appear in the work of Thom as elementary catastrophes (cf. [Th] Chap. 5).

3) The real or complex function germs corresponding to all equations above (up to the addition of a nondegenerate quadratic form in independent variables) have been characterized by Arnol'd in his work [A2] as being the simple singularities in his sense of the term, i.e. as those function germs whose semiuniversal unfoldings (or deformations, as is verified by the later classification results [A3]) produce only finitely many different isomorphism types of neighboring singularities.

4) The "rigidity" of the simple singularities found by Arnol'd parallels the rigidity of two other structures described by Dynkin diagrams. Considering the variety of possible structure constants for a Lie algebra of fixed dimension, the semisimple Lie algebras, if they exist, form an open set which will possibly decompose into finitely many isomorphism components ("semisimple" $\Leftrightarrow$ "Killing form is nondegenerate"). In the theory of the representations of quivers (Gabriel) the "stable" quivers, the representations of which can be arranged into discrete classes of isomorphism types, are likewise given by the homogeneous Dynkin diagrams ([Ga]).

6.2 Symmetries on Rational Double Points.

In this section we consider certain symmetries on rational double points which will play a crucial rôle in the understanding of the subregular deformations in Lie groups and Lie algebras of type B_r, C_r, F_4 or G_2.

We start with some formal constructions. Let Δ be an inhomogeneous irreducible Dynkin diagram. Then its <u>associated homogeneous Dynkin diagram</u> ${}_h\Delta$ is given by the rule

$$ {}_hB_r = A_{2r-1} , \quad {}_hC_r = D_{r+1} , \quad {}_hF_4 = E_6 , \quad {}_hG_2 = D_4 . $$

We define the <u>associated symmetry group</u> $AS(\Delta)$ of Δ by

$$ AS(\Delta) = \begin{cases} \mathfrak{S}_3 & \text{if } \Delta = G_2 \\ \mathbb{Z}/2\mathbb{Z} & \text{else} \end{cases} $$

There is a unique (in case $\Delta = C_3$ or G_2 : up to conjugation by $\mathrm{Aut}({}_h\Delta) = \mathfrak{S}_3$) faithful action of $AS(\Delta)$ on the associated homogeneous diagram ${}_h\Delta$, which we call the associated action of $AS(\Delta)$ on ${}_h\Delta$. In a symbolic way we may regard Δ as the quotient of ${}_h\Delta$ by the associated action of $AS(\Delta)$:

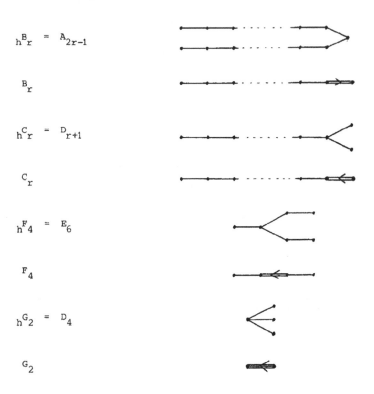

${}_hB_r = A_{2r-1}$

B_r

${}_hC_r = D_{r+1}$

C_r

${}_hF_4 = E_6$

F_4

${}_hG_2 = D_4$

G_2

According to 6.1 a rational double point may be represented as the quotient $\mathbb{A}^2/F$ of $\mathbb{A}^2$ by a finite subgroup F of SL_2 provided char(k) is good (with respect to the corresponding Dynkin diagram). Let F' be another finite subgroup of SL_2 containing F as a normal subgroup. Then the quotient $\Gamma = F'/F$ will act naturally on $\mathbb{A}^2/F$.

Definition: Let Δ be an inhomogeneous irreducible Dynkin diagram and let char(k) be good for Δ . A couple (X,Γ) consisting of a normal surface singularity X and a group Γ of automorphisms of X is called a simple singularity of type Δ if it is isomorphic (after Henselization) to a couple $(\mathbb{A}^2/F\ ,\ F'/F)$ according to the following list:

Δ	F	F'
$B_r\ ,\ r \geq 2$	$\mathbb{Z}_{2r}$	$\mathbb{D}_r$
$C_r\ ,\ r \geq 3$	$\mathbb{D}_{r-1}$	$\mathbb{D}_{2(r-1)}$
F_4	$\mathbb{T}$	$\mathbb{O}$
G_2	$\mathbb{D}_2$	$\mathbb{O}$

If (X,Γ) is a simple singularity of type Δ, then X is a rational double point of type ${}_h\Delta$ and Γ is isomorphic to $AS(\Delta)$. The action of Γ on X lifts in a unique way to an action of Γ on the minimal resolution X' of X. As Γ fixes the singular point of X the exceptional divisor in X' will be stable under Γ. Thus we get an action of Γ on the dual diagram ${}_h\Delta$ of the exceptional divisor. This action agrees with the associated action of $AS(\Delta)$ on ${}_h\Delta$ as may be seen either by the following description and the explicit resolution of X or by the general results obtained later.

An explicit description of the action of $\Gamma = F'/F$ on $\mathbb{A}^2/F$ is obtained by looking at the action of F' on three fundamental F-invariant polynomials on $\mathbb{A}^2$. We let F and F' have the form as described in 6.1, by u and v we denote corresponding coordinates on $\mathbb{A}^2$:

1) B_r. Fundamental invariants under $F = \mathbb{Z}_{2r}$ are $X = uv$, $Y = u^{2r}$, and $Z = v^{2r}$ with the (unique, nontrivial) relation $X^{2r} = YZ$. The element $\begin{pmatrix} 0 & 1 \\ -1 & 0 \end{pmatrix}$ of $\mathbb{D}_r$ operates by $X,Y,Z \mapsto -X,Z,Y$.

2) C_{r+1}, $r \geq 2$. The fundamental invariants of $F = \mathbb{D}_r$ are $X = u^2v^2$, $Y = u^{2r} + v^{2r}$ and $Z = uv(u^{2r} - v^{2r})$ with the relation $Z^2 = XY^2 - 4X^{r+1}$. The quotient $\Gamma = \mathbb{D}_{2r}/\mathbb{D}_r \cong \mathbb{Z}_{4r}/\mathbb{Z}_{2r}$ operates by $X,Y,Z \mapsto X,-Y,-Z$.

3) G_2. We choose as fundamental invariants X,Y,Z under $F = \mathbb{D}_2$ with the relation $X^3 + Y^3 + Z^2 = 0$ the functions (in Klein's notation, see [Kl]) $-\Phi,\Psi,$ and

$\sqrt[4]{-432}\ t$, where

$$\Phi = u^4 + 2\sqrt{-3}\ u^2v^2 + v^4$$
$$\Psi = u^4 - 2\sqrt{-3}\ u^2v^2 + v^4$$
$$t = uv(u^4 - v^4) .$$

The quotient $\mathbb{T}/\mathbb{D}_2 \cong \mathbb{Z}/3\mathbb{Z}$ operates by $X,Y,Z \mapsto \alpha X,\alpha^2 Y,Z$ with $\alpha^3 = 1$, $\alpha \neq 1$, and the involution in $\mathbb{O}/\mathbb{D}_2$ induced by $\begin{pmatrix} \varepsilon & 0 \\ 0 & \varepsilon^7 \end{pmatrix}$ operates by $X,Y,Z \mapsto Y,X,-Z$.

4) F_4 . We choose (again in Klein's notation) as fundamental invariants X,Y,Z under $F = \mathbb{T}$ with the relation $X^4 + Y^3 + Z^2 = 0$ the functions $\sqrt[4]{108}\ t$, $-W$ and χ , where t is as above, $W = \Phi\ \Psi$ (Φ and Ψ as above) and $\chi = u^{12} - 33u^8v^4 - 33u^4v^8 + v^{12}$.

The quotient $\mathbb{O}/\mathbb{T} \cong \mathbb{Z}/2\mathbb{Z}$ operates by $X,Y,Z \mapsto -X,Y,-Z$.

<u>Remark:</u> Klein gives in [Kl] Abschnitt I, Kap. III different fundamental (absolute) invariants with different relations, which however can easily be converted into those above. In the case $k = \mathbb{C}$ the zeroes of the homogeneous polynomials Φ,Ψ,t,W resp. χ on $P^1(\mathbb{C}) \cong S^2$ correspond to the vertices of the regular tetrahedron, its dual tetrahedron, the octahedron, the cube resp. the midpoints of the edges of the octahedron, when these solids are in a certain normal position (cf. [Kl]).

Let (X,x) be an isolated singularity with singular point x , and let Γ be a group of automorphisms of X . Then x will be a fixed point under Γ . For the sake of brevity we call the action of Γ on X <u>free</u> if it is free on the complement $X\setminus x$ of the singular point.

Now let $F' \subset SL_2$ be a finite subgroup and assume char(k) does not divide the order of F' . Since F' contains no reflections it will operate freely on $\mathbb{A}^2\setminus 0$. This again implies that, for a normal subgroup $F \lhd F'$, the group F'/F operates freely on the quotient $\mathbb{A}^2/F$ (for the cases considered by us this can also be seen by the explicit formulae above).

In the rest of this section we will establish the following criterion.

Proposition: Let Δ be a Dynkin diagram of type B_r, C_r, F_4, G_2 and char(k) good for Δ. Let X be a normal surface singularity and Γ a group of automorphisms of X. Assume the following properties hold:

i) X is a rational double point of type ${}_h\Delta$.

ii) Γ is isomorphic to $AS(\Delta)$.

iii) The action of Γ on the dual diagram of the minimal resolution of X coincides with the associated action of $AS(\Delta)$ on ${}_h\Delta$.

iv) Γ acts freely on X.

Then (X,Γ) is a simple singularity of type Δ.

This statement will follow from three auxiliary results.

Lemma 1: The group of bundle automorphisms of the cotangent bundle $T^*\mathbb{P}$ of the projective line $\mathbb{P}$ is isomorphic to $PGL_2 \times G_m$. Here PGL_2 acts on $T^*\mathbb{P}$ by the natural automorphisms induced by the action of $PGL_2 = Aut(\mathbb{P})$ on $\mathbb{P}$, and G_m operates by scalar multiplication on the fibers.

Proof: Let ϕ be a bundle aumorphism of $T^*\mathbb{P}$ and $\bar{\phi} \in PGL_2$ the associated automorphism on the base $\mathbb{P}$. If $\psi : T^*\mathbb{P} \to T^*\mathbb{P}$ is the automorphism naturally induced by $\bar{\phi}$ on $T^*\mathbb{P}$, then $\phi \circ \psi^{-1}$ operates trivially on the base $\mathbb{P}$ and corresponds to a nowhere vanishing section of the bundle $End_{\mathbb{P}}(T^*\mathbb{P})$ of linear endomorphisms of $T^*\mathbb{P}$ over $\mathbb{P}$. Since $T^*\mathbb{P}$ is a line bundle, $End_{\mathbb{P}}(T^*\mathbb{P}) \cong T\mathbb{P} \otimes T^*\mathbb{P} \cong \mathcal{O}$ is trivial, and $\phi \circ \psi^{-1}$ corresponds to a non-zero scalar multiplication.

By means of this lemma it is easy to classify the involutions on $T^*\mathbb{P}$. Up to conjugation there are four different homomorphisms

$$\mathbb{Z}/2\mathbb{Z} \to PGL_2 \times G_m$$

to distinguish:

1) Both projections to the factors are trivial. That will be the trivial action.

2) The projection to PGL_2 is trivial, but the projection to G_m is not. Then the involution operates by -1 on the fibers of $T^*\mathbb{P} \cong \mathcal{O}(-2)$. The quotient by this action is the bundle $\mathcal{O}(-4)$.

3) The projection to PGL_2 is not trivial, but the projection to G_m is. The corresponding involution has two fixed points 0 and ∞ on $\mathbb{P}$. In local coordinates around those points the involution operates by $(x,y) \mapsto (-x,-y)$ (in the fiber direction contragradiently to the horizontal base direction). The quotient has two singularities of type A_1 , which can be resolved into two rational curves with self-intersection number -2 :

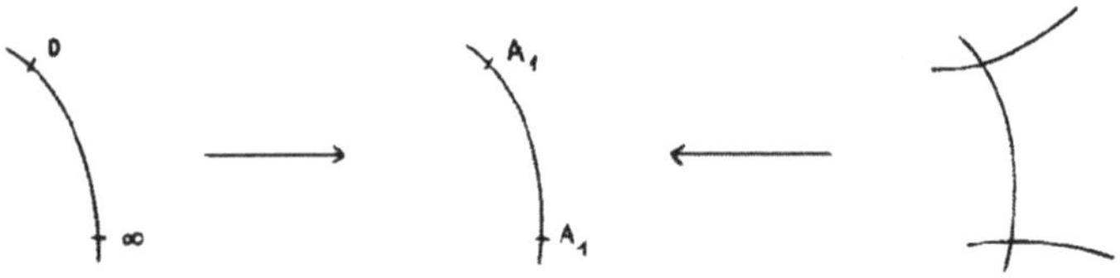

The self-intersection number of the third curve can be gotten from the following interpretation of the situation. The preimage of $\mathbb{Z}/2\mathbb{Z} \subset PGL_2$ in SL_2 is $\mathbb{Z}_4$, and the involution on $T^*\mathbb{P}$ is induced by the action of $\mathbb{Z}_4/\mathbb{Z}_2$ on the singularity $\mathbb{A}^2/\mathbb{Z}_2$, which is resolved by $T^*\mathbb{P}$. The three curves in the resolution of the quotient of $T^*\mathbb{P}$ by the involution will thus resolve the A_3-singularity $(\mathbb{A}^2/\mathbb{Z}_2)/(\mathbb{Z}_4/\mathbb{Z}_2) = \mathbb{A}^2/\mathbb{Z}_4$ and therefore all have self-intersection number -2 .

4) Both projections are not trivial. Following the description of case 3, the fixed points of the involution will consist of the fibers of $T^*\mathbb{P}$ over the points 0 and ∞ . The quotient will be the bundle $\mathcal{O}(-1)$.

<u>Lemma 2</u>: <u>Let</u> Δ <u>be a diagram of type</u> B_r , C_r , F_4 , G_2 <u>and</u> $\mathrm{char}(k)$ <u>good for</u> Δ. <u>Let</u> $AS(\Delta)$ <u>act freely on a rational double point</u> X_0 <u>of type</u> ${}_h\Delta$ <u>in such a way</u>

that the induced action on the dual diagram ${}_h\Delta$ *of the minimal resolution* Y_o *of* X_o *is the associated action of* $AS(\Delta)$. *Then the quotient* X_1 *of* X_o *by the action of* $AS(\Delta)$ *is a rational double point whose type is given by the following list:*

Δ	X_o	X_1
B_r	A_{2r-1}	D_{r+2}
C_r	D_{r+1}	D_{2r}
F_4	E_6	E_7
G_2	D_4	E_7

Proof: We proceed as follows. First we lift the $AS(\Delta)$-action on X_o to its minimal resolution Y_o and form the quotient Y of Y_o by $AS(\Delta)$. This quotient gives in a natural way a partial resolution of the singularities of X_1 . The minimal resolution $Y_1 \to Y$ of the singularities of Y induces a complete resolution of X_1 , and the exceptional divisor of $Y_1 \to X_1$ will allow us to identify X_1 (as a quotient space of a normal space X_1 is also normal, cf. [M2]).

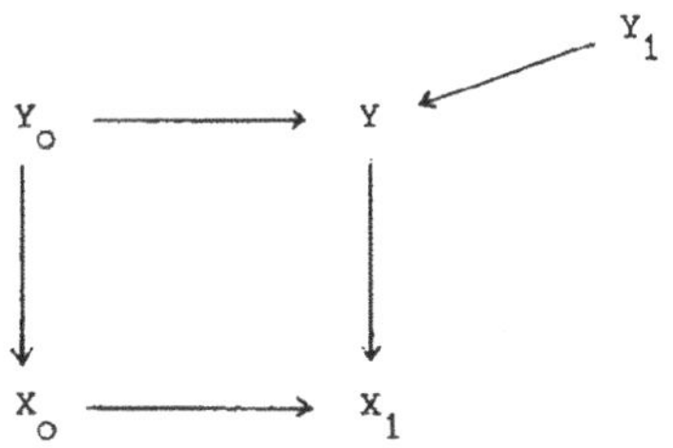

Since $AS(\Delta)$ acts freely on X_o , the singularities of Y can only lie in the image of the exceptional divisor of Y_o under $Y_o \to Y$. To study these singularities up to Hensilization it therefore suffices to consider an étale (or even a formal) neighborhood of the exceptional curves in Y_o . Furthermore we can replace the $AS(\Delta)$-action in the neighborhood of an $AS(\Delta)$-stable exceptional curve by the corresponding linear action on the normal bundle of this curve in Y_o , which is isomorphic to the cotangent bundle $T^*\mathbb{P}$ of the projective line $\mathbb{P} = \mathbb{P}^1$ (a linearization can be carried out for example using the bundle version of the formula from [Br2] Lemma 2).

We now discuss the individual cases:

B_r , $r \geq 2$. Here X_o is of type A_{2r-1} . In the resolution Y_o of X_o , the group $AS(\Delta)$ permutes the "arms" of the exceptional divisor. Since $AS(\Delta)$ acts freely on X_o , the action of $AS(\Delta)$ near the central exceptional component, which is stable under $AS(\Delta)$, must be of type 3 in the preceding discussion. Hence $Y_1 \to X_1$ is a resolution of a singularity of type D_{r+2} :

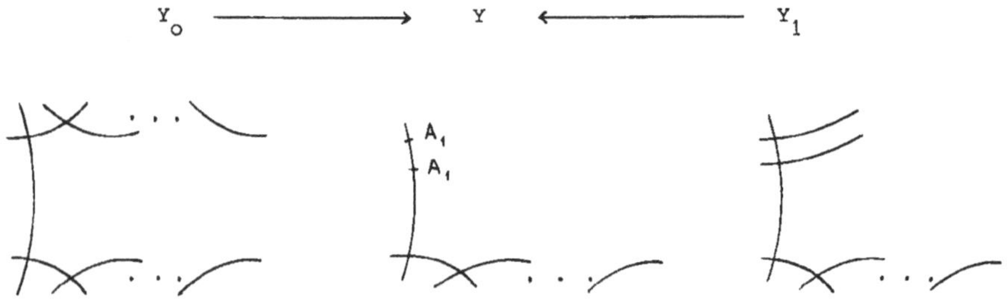

C_r , $r \geq 3$. Here X_o has type D_{r+1} , and $AS(\Delta)$ permutes the (if $r = 3$, two of the) short arms of the exceptional divisor. Since the action of $AS(\Delta)$ on the complement of the exceptional divisor is free, $AS(\Delta)$ must necessarily act on the "branching component" by an involution of type 3. Again on the remaining curves of the long arm we obtain involutions of type 3. The singularity of X_1 is of type D_{2r} :

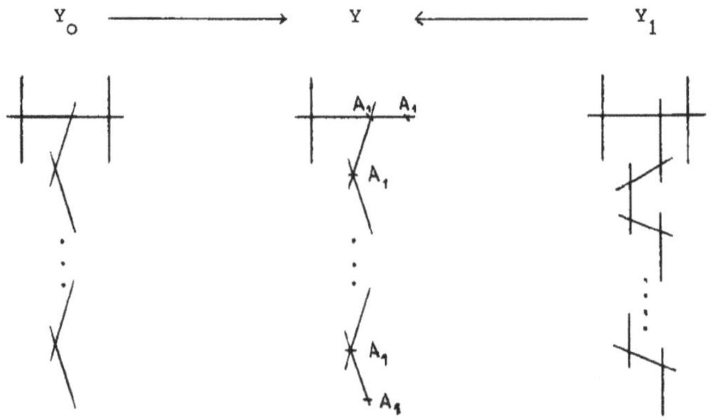

(We note that the involution for this case is already determined by the freeness property alone).

F_4 . For this case, use arguments similar to C_r . For X_1 we get E_7 :

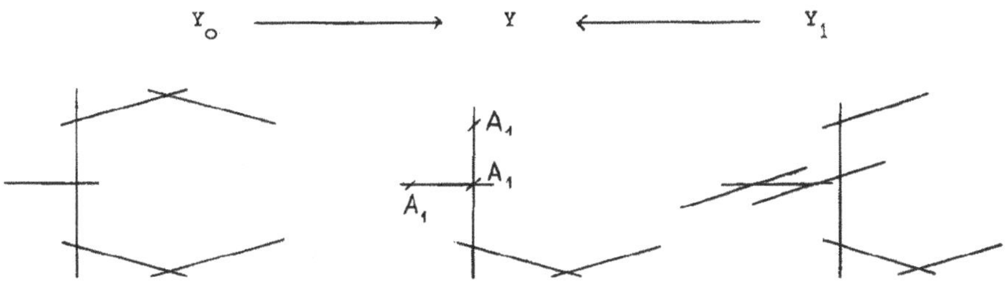

(Here also the action is already determined by the freeness property.)

G_2 . In this case X_o is of type D_4 and $AS(\Delta) \cong \mathfrak{S}_3$ permutes the three arms of the exceptional divisor of Y_o . The action on the central component is given by a homomorphism

$$\mathfrak{S}_3 \to \mathrm{Aut}(T^* \mathbb{P}) = PGL_2 \times G_m ,$$

whose first projection is faithful and is determined uniquely up to conjugation. The restriction of the second projection to $\mathfrak{A}_3 \cong \mathbb{Z}/3\mathbb{Z}$ is automatically trivial. The action of $\mathfrak{A}_3$ on the central component then has two fixed points 0 and ∞ . In local coordinates around these points the action of $\mathfrak{A}_3$ has the form

$$(x,y) \mapsto (\zeta x, \zeta^{-1} y) , \quad \zeta^3 = 1$$

(fiber contragradient to the base). Therefore in the quotient of Y_o by $\mathfrak{A}_3$ we get two singularities of type A_2 , each of which can be resolved into two rational curves with self-intersection number -2 :

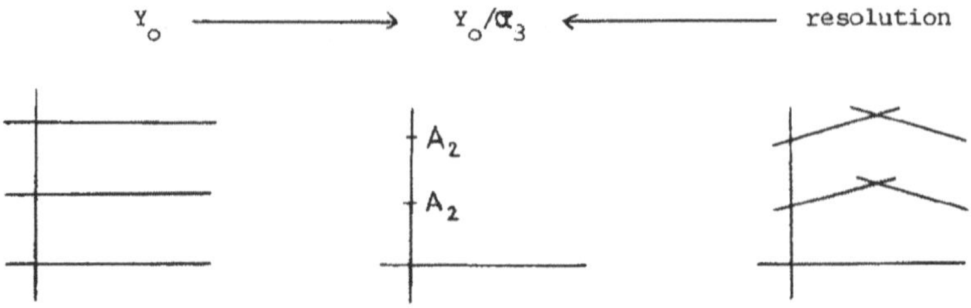

For the self-intersection of the preimage of the central curve we obtain -2 by a similar reasoning as in case 3 of the $\mathbb{Z}/2\mathbb{Z}$-actions. That is, the group $\mathfrak{A}_3 \subset PGL_2$ can be lifted to $\mathbb{Z}_6$ in SL_2, and its action on $T^*\mathbb{P}$ is induced by the action of the quotient group $\mathbb{Z}_6/\mathbb{Z}_2$ on the quotient $\mathbb{A}^2/\mathbb{Z}_2$ of type A_1, which is resolved by $T^*\mathbb{P}$. The quotient $(\mathbb{A}^2/\mathbb{Z}_2)/(\mathbb{Z}_6/\mathbb{Z}_2) = \mathbb{A}^2/\mathbb{Z}_6$ is a singularity of type A_5 and is minimally resolved by the resolution of $T^*\mathbb{P}/\mathfrak{A}_3$. With that the quotient $X_o/\mathfrak{A}_3$ has a singularity of type E_6, on which the group $\mathfrak{S}_3/\mathfrak{A}_3 = \mathbb{Z}/2\mathbb{Z}$ operates freely. As in the case F_4 (closing remark), the quotient $X_1 = X_o/\mathfrak{S}_3 = E_6/(\mathfrak{S}_3/\mathfrak{A}_3)$ has a singularity of type E_7.

Lemma 3: Let X_o be a rational double point and Γ a finite group of freely acting automorphisms of X_o. Assume that the quotient $X_1 = X_o/\Gamma$ is again a rational double point and that char(k) is good for X_1. Then there is a finite subgroup F' of SL_2 and a normal subgroup F of F' such that (after Henselization) (X_o,Γ) resp. X_1 is isomorphic to $(\mathbb{A}^2/F, F'/F)$ resp. $\mathbb{A}^2/F'$.

Proof: We consider the Henselizations $\tilde{\mathbb{A}}^2$ of $\mathbb{A}^2$ at 0, and $\tilde{X}_o, \tilde{X}_1$ of X_o, X_1 at their singular points. By U, V_o, V_1 we denote the complements of the closed points in $\tilde{\mathbb{A}}^2, \tilde{X}_o, \tilde{X}_1$. Under our assumptions we may regard $\tilde{X}_1$ as the quotient of $\tilde{\mathbb{A}}^2$ by a finite subgroup F' of SL_2 acting freely on U. Since U is simply connected F' may be identified with the fundamental group of V_1 (cf. [Ar5] or [Br2] in the complex analytic case). The quotient $\tilde{X}_o \to \tilde{X}_o/\Gamma = \tilde{X}_1$ induces a connected étale covering $V_o \longrightarrow V_1$ with Galois group Γ. Thus there exists a normal subgroup F of

F' such that $V_o \cong U/F$ and $(V_o,\Gamma) \cong (V_o,F'/F)$. Since $\tilde{X}_o$ and $\tilde{X}_1$ are normal we obtain $\tilde{X}_o = \tilde{\mathbb{A}}^2/F$ and $(\tilde{X}_o,\Gamma) \cong (\tilde{\mathbb{A}}^2/F,F'/F)$. This proves the lemma.

Lemma 2 and Lemma 3 now imply the proposition.

Remarks: 1) In the proof of Lemma 2 we already remarked that for Δ of type C_r , F_4 condition iii) of the Proposition is already implied by the other three conditions. This is also true for G_2 but not for B_r . In the last case we can even find two nonequivalent involutions on the corresponding singularity of type A_{2r-1} which act freely and induce the trivial action on the resolution diagram (i.e. those induced by the inclusions $\mathbb{Z}_m \subset \mathbb{Z}_{2m}$ in SL_2 and $\mathbb{Z}_m \subset \left\{ \begin{pmatrix} \alpha & \\ & \alpha^{m-1} \end{pmatrix} \mid \alpha \in \mu_{2m} \right\}$ in GL_2 , where $m = 2r$).

2) After the appearence of the first version of this work Arnol'd independently introduced singularities of type B_r , C_r , F_4 in a different but related sense. He studies critical points of differentiable functions on manifolds with boundary. Such functions correspond to functions on the double manifolds (without boundary) invariant with respect to the canonical involution (cf. [A4]).

6.3. Dynkin Curves.

Let G be a simple, simply connected group, B a Borel subgroup of G with unipotent radical U , and $\pi : G \times^B U \to V(G)$ the resolution of the unipotent variety of G , as in 4.1. In preparing for the identification of the subregular singularities of G in the next section, 6.4, we now look at the fiber $\pi^{-1}(x)$ of π over a subregular unipotent element $x \in V(G)$. As in 4.1, we factor π through the embedding τ

$$\begin{array}{ccccc} G \times^B U & \overset{\tau}{\hookrightarrow} & G/B \times V(G) & \xrightarrow{p_2} & V(G) \\ g * u & \longmapsto & (gB\ ,\ {}^g u) & \longmapsto & {}^g u\ . \end{array}$$

The image $\tau(\pi^{-1}(x)_{red})$ of the reduced fiber of π over x has the form $\{(B',x) \in \mathcal{B} \times \{x\} \mid x \in B'\} \cong \mathcal{B}_x := \{B' \in \mathcal{B} \mid x \in B'\}$ if we identify G/B with the variety $\mathbb{B}$ of all Borel subgroups of G .

Let Δ be that system of simple roots of the maximal torus $T \subset B$ in G which is uniquely determined by B . For every $\alpha \in \Delta$ let P_α denote the parabolic subgroup of G generated by B and $U_{-\alpha}$ (cf. 3.1). The fibers of the natural projection $G/B \to G/P_\alpha$ have the form gP_α/B , $g \in G$, and are isomorphic to the projective line (cf. [St2] 3.9, Prop. 1, Proof).

<u>Definition:</u> Such a fiber (as described above) is said to be a <u>line of type</u> α <u>in</u> G/B . Two lines of types α and $\beta \neq \alpha$ intersect in at most one point ([St2] 3.10, Prop. 1, c) and with different tangents (if the lines meet, without loss of generality, at $B \in G/B$, then in the local chart $U^- \cdot B$ at B in G/B , which is isomorphic to U^- , these lines have the form $U_{-\alpha}$ resp. $U_{-\beta}$).

Let $((n_{\alpha,\beta}))_{\alpha,\beta \in \Delta}$ be the Cartan matrix of Δ . Then let

$$n'_{\alpha,\beta} := \begin{cases} -n_{\alpha,\beta} & \text{when } n_{\alpha,\beta} = 0 \text{ or } \operatorname{char}(k) \nmid n_{\alpha,\beta} \\ 1 & \text{else.} \end{cases}$$

A look at the simple root systems shows that $n_{\alpha,\beta}$ is not divisible by char(k) when the characteristic is good, so that in this case the matrix $((n'_{\alpha,\beta}))$ is just the negative Cartan matrix.

<u>Definition:</u> A <u>Dynkin curve</u> in G/B is a nonempty union E of lines of different types $\alpha \in \Delta$ with the properties

i) E is connected,

ii) every line of type α in E intersects exactly $n'_{\alpha,\beta}$ lines of type β for $\alpha \neq \beta$.

Let the <u>dual diagram of a Dynkin curve</u> be that diagram in which the lines of the curve are represented by the vertices, and two vertices are connected when the correponding lines intersect.

If Δ is a homogeneous (resp. inhomogeneous) irreducible Dynkin diagram, and G is simple of type Δ, then the dual diagram of a Dynkin curve in G/B is of type Δ (resp. ${}_h\Delta$, when char(k) is good).

Example: Let $\Delta = G_2$, char(k) $\neq 3$. The Cartan matrix has the form $\begin{pmatrix} 2 & -1 \\ -3 & 2 \end{pmatrix}$. A Dynkin curve consists of one line of type α and three lines of type β, i.e. $E = C_\alpha \cup C_\beta \cup C'_\beta \cup C''_\beta$, where β is the longest root, with the following intersection behavior:

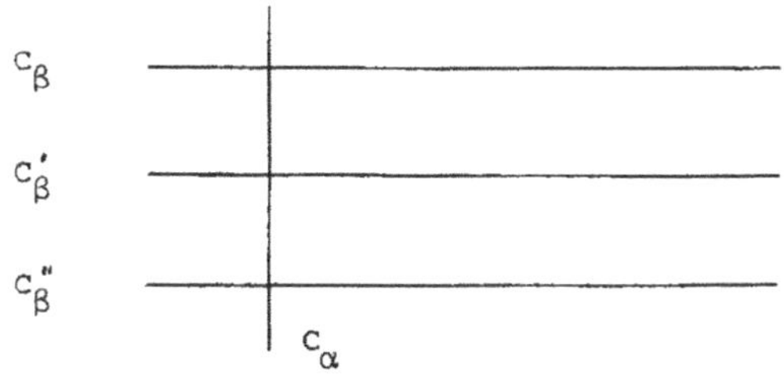

The associated homogeneous diagram is ${}_hG_2 = D_4$.

The following theorem is due to Steinberg and Tits. A proof is found in [St2] 3.10, Th. 3 and Prop. 3.

Theorem: A unipotent element x of the simple group G is subregular exactly when the reduced fiber $\pi^{-1}(x)_{red} \cong \mathbb{B}_x$ is a Dynkin curve in $\mathbb{B} = G/B$. All Dynkin curves in $\mathbb{B}$ are conjugate, in particular every Dynkin curve occurs as a fiber $\mathbb{B}_x$ of π over some appropriate subregular unipotent element $x \in G$.

Example: In the case A_n, the above description of the fiber $\mathbb{B}_x$ can be realized in an elementary way. A subregular element x of the group SL_{n+1} can be represented in the form

$$\left(\begin{array}{c|ccccc} 1 & & & 0 & & \\ \hline & 1 & 1 & & & 0 \\ & & 1 & \ddots & & \\ & & & \ddots & \ddots & \\ 0 & 0 & & & 1 & 1 \\ & & & & & 1 \end{array}\right)$$

by choosing an appropriate basis $e_o, e_1, \ldots, e_n$ of $\mathbb{A}^{n+1}$ (cf. [St2] pp 135, 136). Under the map $F \in \mathbb{F} \mapsto$ (stabilizer of F in SL_{n+1}) the flag manifold $\mathbb{F}$ of complete flags $F = (F_1, \ldots, F_n)$ of vector subspaces $F_i \subset \mathbb{A}^{n+1}$, $\dim F_i = i$, $F_i \subset F_{i+1}$, can be identified with the variety $\mathbb{B}$ (cf. [Bo] III 10.3). The fiber $\mathbb{B}_x$ corresponds to the variety $\mathbb{F}_x = \{F \in \mathbb{F} \mid xF = F\}$ of all flags stabilized by x.

For $j = 1, \ldots, n$ and $(\lambda:\mu) \in \mathbb{P}^1$ let the flag $F^j(\lambda:\mu)$ be defined as follows (with $\langle v_1, \ldots, v_i \rangle$ denoting the vector space generated by $v_1, \ldots, v_i \in \mathbb{A}^{n+1}$) :

$$F_i^j(\lambda:\mu) := \begin{cases} \langle e_1, \ldots, e_i \rangle & \text{for } i < j \\ \langle e_1, \ldots, e_{i-1}, \lambda e_i + \mu e_o \rangle & \text{for } i = j \\ \langle e_o, e_1, \ldots, e_{i-1} \rangle & \text{for } i > j \end{cases}$$

The flags of the n one-parameter families $F^j = \{F^j(\lambda:\mu) \mid (\lambda:\mu) \in \mathbb{P}^1\}$, $j=1,\ldots,n$ are exactly those which are stabilized by x. Two different families F^j and F^k intersect in at most one point, and only when $|j-k| = 1$, i.e. $F^{j-1}(\lambda:0) = F^j(0:\mu)$ is the single intersection of the flags. This gives a configuration for $\mathbb{F}_x$ of a Dynkin curve of type A_n :

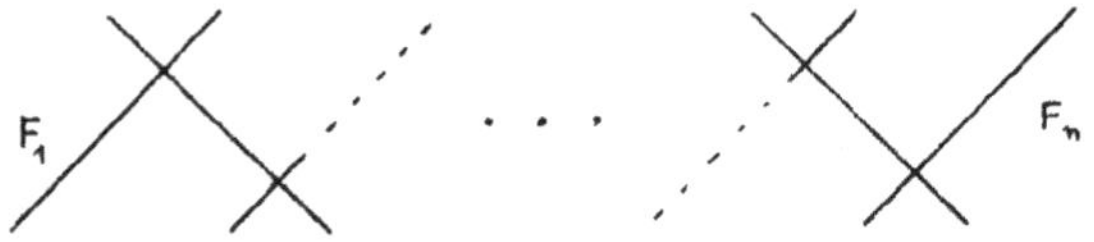

6.4. The Identification of the Subregular Singularities.

Let G be a simple group of type Δ, char(k) good for G, S a transverse slice to the subregular unipotent orbit of G, and $\sigma : S \to T/W$ the restriction of the adjoint quotient to S as in 5.2, 5.3 and 5.5. Let $X = \sigma^{-1}(\bar{e}) = S \cap V(G)$ be the intersection of S with the unipotent variety $V(G)$, and let $x \in G$ be a singular, hence subregular, point of X. As we saw in 5.3, the restriction $\pi_{|X'} : X' \to X$ of the resolution $\pi : G \times^B U \to V(G) \subset G$ to the preimage $X' = \pi^{-1}(X) = \pi^{-1}(S)$ of S in $G \times^B U$ is a resolution of X, which for simplicity we will also call π

(in the notation of 5.3 we have $X' = \Theta'^{-1}(e))$. By the results of 6.3, the resolution $\pi : X' \to X$ will have an exceptional configuration of type Δ over x when the self-intersection numbers of the reduced components of $\pi^{-1}(x)$ are equal to -2 (cf. 6.1). We will derive that now, and so identify the singularity of X at x as a rational double point of type Δ .

Lemma: Let G be a simple group of rank 1 , $T \subset B$ a maximal torus T contained in a Borel subgroup B of G , and let α be the root of T in B . Let F be an affine line on which B operates by means of the character $B \to T \xrightarrow{\alpha} G_m$ induced by α . Then the first Chern class of the associated bundle $G \times^B F$ over $G/B \cong \mathbb{P}^1$ is equal to -2 .

Proof: One easily sees that it suffices to discuss the case $G = SL_2$. Since the formation of associated fiber bundles commutes with tensor products, it suffices to prove that the bundle $\mathcal{O}(-1)$ over $\mathbb{P}^1$ is isomorphic to the bundle $G \times^B F_\omega$, where B operates on the affine line F_ω by the weight $\omega = \alpha/2$. We identify G/B with $\mathbb{P}^1$ in the following way: The class B in G/B shall correspond to the unique line $D \subset \mathbb{A}^2$ stabilized by B in the natural representation of $G = SL_2$ on $\mathbb{A}^2$. As B operates on D by the weight ω , we may identify D with F_ω ; the class gB then corresponds to the line $gF_\omega \subset \mathbb{A}^2$. The embedding $G \times^B F_\omega \to \mathbb{P}^1 \times \mathbb{A}^2$, $g * f \mapsto (gF_\omega, gf)$ gives an isomorphism (cf. 3.7) onto the tautological bundle $L = \{(F,f) \in \mathbb{P}^1 \times \mathbb{A}^2 \mid f \in F\}$, that is known to be isomorphic to $\mathcal{O}(-1)$ (for example, consider the meromorphic section $s : \mathbb{P}^1 \to L$, $s(\lambda:\mu) = ((\lambda:\mu),(\lambda/\mu,1))$, which has exactly one simple pole at $\mu = 0$).

For the next theorem, we recall briefly the structure of the parabolic subgroups P_α . which are generated by a Borel subgroup $B \subset G$ and a root subgroup $U_{-\alpha}$, $\alpha \in \Delta$ (here Δ is the system of simple roots determined by a maximal torus $T \subset B$ and B). For brevity, denote P_α by P (for details in the following cf. [B-T] 4., [Hu] 30.2). The solvable resp. unipotent radical $R(P)$ resp. $R_u(P) = U_P$ of P is contained in B resp. the unipotent radical U of B . The quotient $P/R(P)$ is a simple group of rank 1. The group U_P has the form $U_P = \prod_{\beta \in \Sigma^+ \setminus \{\alpha\}} U_\beta$, where Σ^+ denotes the set of positive roots of T in G. The inclusion $U_\alpha \hookrightarrow U$ composed with the projection $U \to U/U_P$ gives an isomorphism $U_\alpha \xrightarrow{\sim} U/U_P$. The group P normalizes U_P, so $B \subset P$ normalizes U_P too, and hence B operates on the quotient U/U_P. The commutator formula ([St2] 3.7, p. 111; [Bo] IV, 14.5, pp. 334,335) shows that U operates trivially on U/U_P. Therefore B operates on U/U_P by the root α, i.e. via the composition $B \to T \xrightarrow{\alpha} G_m \cong \mathrm{Aut}(U_\alpha) \cong \mathrm{Aut}(U/U_P)$. The maximal torus $T \cap R(P)$ of $R(P)$ lies in the center of the Levi subgroup L of P. As L is generated by T, U_α, $U_{-\alpha}$, the torus $T \cap R(P)$ lies in the kernel of the root α. Hence $R(P)$ also acts trivially on U/U_P because $R(P) = U_P \cdot (T \cap R(P))$.

Using the original notations we have:

<u>Proposition</u>: <u>The components of the reduced exceptional curve</u> $\pi^{-1}(x)_{red}$ <u>in</u> X' <u>have self-intersection number</u> -2.

<u>Proof</u>: Let $C \cong \mathbb{P}^1$ be a component of $\pi^{-1}(x)_{red}$. We will show that the first Chern class of the normal bundle $N_{C|X'}$ of C in X' equals the first Chern class of the normal bundle $N_{C|G \times^B U}$ of C in $G \times^B U$, and that the latter has value -2.

The inclusions $C \subset X' \subset G \times^B U$ are immersions of smooth manifolds and by EGA IV 19.1.5 there is an exact sequence for the corresponding normal bundles

$$0 \to N_{C|X'} \to N_{C|G \times^B U} \to i^* N_{X'|G \times^B U} \to 0$$

(here i denotes the inclusion $C \hookrightarrow X'$). By the functoriality and the additivity of the first Chern class with respect to exact sequences (cf. [G] 4-17,18), it follows that $c_1(N_{C|X'})$ and $c_1(N_{C|G \times^B U})$ are equal if we can prove that $N_{X'|G \times^B U}$ is trivial. For that, we consider the cartesian diagram

$$\begin{array}{ccc} X' & \xrightarrow{\pi|_{X'}} & S \\ \downarrow & & \downarrow \\ G \times^{B} U & \xrightarrow{\pi} & G \end{array} \quad .$$

Since X' is smooth, it follows from the local statement in EGA IV 17.13.2 that the bundle $N_{X'|G \times^{B} U}$ is isomorphic to the induced bundle $(\pi|_{X'})^{*} N_{S|G}$. However, $N_{S|G}$ is trivial for a sufficiently small chosen slice S since S is a smooth submanifold of G at x (cf. EGA IV 16.9.8 or 19.1.1). Now we must calculate $c_1(N_{C|G \times^{B} U})$.

Let the component C be a line of type $\alpha \in \Delta$. Then its image under the embedding $G \times^{B} U \to G/B \times V(G)$, $g * u \mapsto (gB, {}^{g}u)$, has the form $gP_{\alpha}/B \times \{x\}$ for an appropriate $g \in G$. By conjugation of x with g^{-1} we may assume that $gP_{\alpha} = P_{\alpha}$ and $x \in U_{P_{\alpha}} = R_u(P_{\alpha})$ (cf. [St2] 3.10, Prop. 1 (b)). For simplicity let P denote P_{α} .

The restriction of the bundle $G \times^{B} U$ to the fiber P/B of the natural projection $G/B \to G/P$ is the bundle $P \times^{B} U$. The curve C is then a section of $P \times^{B} U$. Since $P \times^{B} U$ is the fiber of the smooth morphism $G \times^{B} U \to G/B \to G/P$, it has a trivial normal bundle $N_{P \times^{B} U|G \times^{B} U}$. The exact sequence

$$0 \to N_{C|P \times^{B} U} \to N_{C|G \times^{B} U} \to j^{*} N_{P \times^{B} U|G \times^{B} U} \to 0$$

where j denotes the inclusion $C \hookrightarrow P \times^{B} U$, reduces our calculation to that of $c_1(N_{C|P \times^{B} U})$. Since $P \times^{B} U$ is a group bundle, the normal bundle of a section can be identified with the normal bundle of the identity section, i.e. with the bundle $P \times^{B} \underline{u}$, where $\underline{u}$ is the Lie algebra of U on which B operates by means of the adjoint representation (cf. 3.7, Corollary). The unipotent radical U_P as well as its Lie algebra $\underline{u}_P$ are normalized by P and $B \subset P$. The exact sequence

$$0 \to \underline{u}_P \to \underline{u} \to \underline{u}/\underline{u}_P \to 0$$

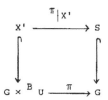

of B-modules induces an exact bundle sequence

$$0 \to P \times^B \underline{u}_P \to P \times^B \underline{u} \to P \times^B (\underline{u}/\underline{u}_P) \to 0 \ .$$

As the B-action on $\underline{u}_P$ is the restriction of the P-action, the bundle $P \times^B \underline{u}_P$ is trivial, i.e. isomorphic to $P/B \times \underline{u}_P$ (3.7 Lemma 1). Since the radical $R(P)$ of P operates trivially on $\underline{u}/\underline{u}_P$, by 3.7 Lemma 3 the bundle $P \times^B (\underline{u}/\underline{u}_P)$ is isomorphic to $P/R(P) \times^{B/R(P)} (\underline{u}/\underline{u}_P)$. Now $P/R(P)$ is simple of rank 1 and its Borel subgroup $B/R(P)$ operates on $\underline{u}/\underline{u}_P$ by the positive root α. By the previous lemma this bundle has first Chern class -2. Since $P \times^B \underline{u}_P$ is trivial, it now follows that $c_1(P \times^B \underline{u}) = -2$, which is what was to be shown.

Remark: The basic idea of the proof above is the same as that of the proof given by Esnault ([Es] III). According to a remark in [Es] it goes back to Deligne. Another way to prove the lemma is also shown in [Es] : by means of an easy calculation the subregular singularity of $V(SL_2)$ at e can be identified as a rational double point of type A_1, which is minimally resolved by the bundle $G \times^B U$ (for $G = SL_2$). Therefore the exceptional identity section of $G \times^B U$ must have self-intersection number -2.

We can now summarize the results of 6.1, 6.3 and 6.4:

Theorem: Let G be a simple, simply connected group with Dynkin diagram Δ, and char(k) good for G. For any unipotent element $x \in G$, the following are equivalent:

i) x is subregular.

ii) The intersection $S \cap V(G)$ of a transverse slice S to the orbit of x with the unipotent variety $V(G)$ has a rational double point at x. The type of this singularity is Δ resp. $_h\Delta$ when Δ is homogeneous resp. inhomogeneous.

Remark: 1) If $G \to G'$ is a separable isogeny of simple groups and G is simply connected, then $V(G)$ and $V(G')$ are isomorphic with respect to the action of the adjoint group. Therefore the above theorem also holds for G' .

2) If char(k) is very good for G , then a result analogous to the Main Theorem can be proven for the Lie algebra $\underline{g}$ of G , either by modifying all the results required for the proof or by using the isomorphism between $V(G)$ and the nilpotent variety $N(\underline{g})$ (cf. Rem. (i) in 3.15).

3) If G is a classical group, then the above theorem can be obtained by direct calculation with the explicitly known invariant polynomials on the Lie algebra ([A1], [He] 4.5.6).

4) Under a somewhat stronger restriction on the characteristic an alternate identification of the subregular singularities will be carried out in 8.3.

5) If the characteristic of k is not good for G , the arguments in the sections above cannot be used directly, because they require the separability of the orbit map of G onto the subregular orbit.

Instead of a slice transverse to the G-action as in 5.1, a slice transverse to the submanifold given by an orbit might be considered, whose dimension equals the codimension of the orbit. For these slices, 5.1 Lemma 3 on the equivalence of different slices is no longer available. We believe that the splitting up of the isomorphism classes of rational double points with equal Dynkin diagram which appears in Artin's classification for bad characteristic is connected with this phenomenon (in every characteristic there exists exactly one subregular orbit, cf. 5.4, so no parallel splitting up of the subregular orbit occurs!).

6.5. The Neighboring Singularities.

Let G , x and $\sigma : S \to T/W$ be as in 6.4. There we identified the singularities of the fiber $\sigma^{-1}(\bar{e})$ as rational double points of the type associated to G . Now we consider the singularities of the other fibers of σ which are also normal and two-dimensional.

For brevity we will use the following conventions. Let the Dynkin diagram of a reductive group be that of its semisimple commutator. If Δ is an inhomogeneous Dynkin diagram, then a rational double point of type Δ will be one of type ${}_h\Delta$. (We reserve the notation "simple singularity of type Δ " for couples (X,Γ) as introduced in 6.2).

Lemma 1: Let $t \in T$, and let $y \in S$ be a singular point of the fiber $\sigma^{-1}(\bar{t})$. If $\Delta(t) = \Delta_1 \cup \ldots \cup \Delta_m$ is the decomposition of the Dynkin diagram of $Z_G(t)$ into connected components, then $\sigma^{-1}(\bar{t})$ has a rational double point at y of type Δ_i for a suitable $i \in \{1,\ldots,m\}$.

Proof: By 5.2 and 5.5, y is a subregular element of G whose G-orbit is transverse to S at y . In particular, $\sigma^{-1}(\bar{t}) = S \cap \chi^{-1}(\bar{t})$ is a transverse slice to the G-orbit of y in $\chi^{-1}(\bar{t})$. To determine the type of the Henselization of $\sigma^{-1}(\bar{t})$ at y , it is enough by 5.1 Lemma 3 to consider a particular transverse slice in $G \times^{Z(t)} V(t)$ to the G-orbit of the image $g * u$ of y under the G-isomorphism $\chi^{-1}(\bar{t}) \xrightarrow{\sim} G \times^{Z(t)} V(t)$ (3.10 Lemma). Since the bundle $G \times^{Z(t)} V(t)$ is G-homogeneous we need only construct a transverse slice to the Z(t)-orbit of u in $V(t)$. By 5.4, the unipotent variety $V(t)$ is the product $V_1 \cdot \ldots \cdot V_m$ of the unipotent varieties of the simple normal subgroups G_i , $i=1,\ldots,m$, of $Z(t)$. Then u can be written as a product $u_1 \cdot \ldots \cdot u_m$ of elements $u_i \in V_i$, $i=1,\ldots,m$. Without loss of generality we may assume that u_1 is subregular in G_1 and that all the other u_i are regular in G_i , $i=2,\ldots,m$. Finally, if S_1 is a transverse slice to the G_1-orbit of u_1 in V_1 , then because of the G_i-regularity of u_i , $i=2,\ldots,m$, the variety $S_1 \cdot u_2 \cdot \ldots \cdot u_m$ will be a transverse slice to the Z(t)-orbit of u in $V(t)$ whose singularity at u is a rational double point of the type corresponding to G_1 (6.4 Theorem, for positive characteristic note Rem.1 as well as 3.6 and 3.13).

Lemma 2: With the notation of Lemma 1 we have: If t lies in a sufficiently small neighborhood U of e in T, then there exists an injection $\Delta_i \mapsto y_i$, $i=1,\ldots,m$ of the components Δ_i of the Dynkin diagram of $Z_G(t)$ to the set of singular points of $\sigma^{-1}(\bar{t})$ such that $\sigma^{-1}(\bar{t})$ has a rational double point of type Δ_i at y_i.

Proof: We will construct an open neighborhood U' of $\bar{e}$ in T/W such that the slice S intersects all the subregular orbits of the fiber $\chi^{-1}(\bar{t})$ for $\bar{t} \in U'$. The preimage U of U' in T is then the desired neighborhood of $e \in T$, since the subregular G-orbits in $\chi^{-1}(\bar{t})$ correspond to just the subregular $Z(t)$-orbits in $V(t)$, which again correspond to the components Δ_i of $\Delta(t)$.

The image of the smooth morphism $\mu : G \times X \to G$, $\mu(g,s) = {}^g s$, is open in G. So the complement $K = G \setminus \mu(G \times S)$ is closed in G and stable under the adjoint action of G. According to 5.4 and 3.10 the fiber $\chi_K^{-1}(\bar{e}) = K \cap \chi^{-1}(\bar{e}) = K \cap V(G)$ of the restriction χ_K of χ to K consists of finitely many orbits of dimension strictly smaller than $n-r-2$, where $n = \dim G$ and $r = \operatorname{rank} G$. Hence $\chi_K^{-1}(\bar{e})$ has dimension strictly smaller than $n-r-2$. Because of the upper semi-continuity of the fiber dimension of the morphism χ_K (cf. EGA IV 13.1.3), the set $C = \{c \in K \mid$ the fiber $\chi_K^{-1}(\chi(c))$ has a component containing c of dimension $\geq n-r-2\}$ is closed in K and in G. Moreover, the fibers of χ_K are G-varieties, and so C is stable under G. Therefore the image $\chi(C)$ is closed in T/W (cf. [M2] 1, § 2, [Sp3] 2.4.8). The complement U' of $\chi(C)$ will then be open in T/W, and will contain $\bar{e}$. The fibers $\chi_K^{-1}(\bar{t})$, $\bar{t} \in U'$, have dimension $< n-r-2$, and so cannot contain any subregular orbits (who have dimension $n-r-2$). So for $\bar{t} \in U'$, such orbits lie in $\chi^{-1}(\bar{t}) \setminus (K \cap \chi^{-1}(\bar{t})) = \chi^{-1}(\bar{t}) \cap \mu(G \times S)$, i.e. S intersects all the subregular orbits of the fibers $\chi^{-1}(\bar{t})$, $\bar{t} \in U'$, as was to be shown.

By 3.5 there is a W-invariant open neighborhood Q of e in T with the property that for all $t \in Q$, the Dynkin diagram $\Delta(t)$ of $Z_G(t)$ can be identified with a proper subdiagram of the diagram Δ of G. So we have the addition:

Lemma 3: If $\Delta' \subset \Delta$ is a proper subdiagram of Δ, then in every neighborhood Q

<u>of</u> e <u>in</u> T <u>there is an element</u> $t \in Q$ <u>with</u> $\Delta(t) = \Delta'$.

<u>Proof</u>: Let Σ be the root system of T in G , let Σ' be the root subsystem generated in Σ by Δ' , let $Q' = \{t \in Q \mid \alpha(t) = 1$ for all $\alpha \in \Delta'\}$, and let $T_\beta = \{t \in T \mid \beta(t) = 1\}$ for $\beta \in \Sigma$. For all $t \in Q'$ which lie in the open complement of $\bigcup_{\beta \in \Sigma \setminus \Sigma'} T_\beta$, we have $\Delta(t) = \Delta'$, and every neighborhood of e in T intersects this complement.

<u>Remark</u>: The statements in this section also hold for the adjoint representation of G on the Lie algebra $\underline{g}$ (char(k) very good). In this case we are spared the choice of a neighborhood Q (3.5), since the centralizer of a semisimple element will have a $\mathbb{Q}$-closed root subsystem (cf. 3.14).

6.6. <u>Neighboring Fibers</u>.

(This section will be used only in 8.10). We look at the same situation as in 6.5 and consider the possible number of singularities that could appear in a fiber of σ . By the explanations in 6.5, the numbers that could occur are determined by the frequency with which the subregular orbits in a fiber of χ meet the slice S . In order to be able to find any statements which don't depend on the choice of S , we subject S to certain restrictions, whose realizability we will encounter later in the applications (cf. 8.10).

<u>Assumptions</u>: Let the slice S intersect the subregular unipotent orbit of G at exactly one point x , and let the singular locus of $\sigma : S \to T/W$ be finite over a neighborhood N of $\bar{e}$ in T/W .

Instead of the two conditions above it suffices, since $\sigma^{-1}\sigma(x) = \{x\}$, to claim only that the singular locus $\{s \in S \cap \sigma^{-1}(N) \mid s$ is a singular point of $\sigma^{-1}\sigma(s)\}$ is proper over N (EGA II 6.1.1, III 4.4.2, or IV 18.12.4).

<u>Remark</u>: The fact that the singular locus is proper over T/W guarantees that the

singularities of the neighboring fibers are also "genuinely" in the "neighborhood" of the point x. This can also be obtained by looking at the Henselizations of T/W at $\bar{e}$ and of S at x, or over $\mathbb{C}$ by replacing S by a sufficiently small Hausdorff neighborhood around x. However, the usual localization of S at x will not suffice.

Let $B \supset T$ be a Borel subgroup of G containing the maximal torus T and let $\Delta = \{\alpha_1,\ldots,\alpha_r\}$ be the simple roots of T in $B \subset G$. For $i \in \{1,\ldots,r\}$ let P_i, U_i, T_i and A_i denote the parabolic subgroup of G generated by $U_{-\alpha_i}$ and B, its unipotent radical, the subgroup $\{t \in T \mid \alpha_i(t) = 1\}$ and the semidirect product $T_i \ltimes U_i$ in P_i. We consider again the diagram defined in 4.3, 4.6

$$\begin{array}{ccccc} G \times^{P_i} A_i & \xrightarrow{\ \phi_i\ } & D_i & \hookrightarrow & G \\ \Big\downarrow{\scriptstyle \theta_i} & & \Big\downarrow{\scriptstyle \chi_i} & \swarrow{\scriptstyle \chi} & \\ T_i & \xrightarrow{\ \psi_i\ } & T/W & & \end{array}$$

where D_i denotes the image of ϕ_i.

As we saw in 4.6, all irregular orbits of G lie in the union $\bigcup_{i=1}^{r} D_i$, and the orbits of maximal dimension in $\bigcup_{i=1}^{r} D_i$ are just the subregular orbits of G. The transverse slice S intersects only regular and subregular orbits of G. Let $S_i = S \cap D_i$. Thus $\bigcup_{i=1}^{r} S_i = S \cap \bigcup_{i=1}^{r} D_i$ consists of only subregular elements. By 5.2 (see also 6.5) this union is just the critical locus of σ.

If O is a subregular orbit in a fiber $\chi^{-1}(\bar{t})$, $\bar{t} \in T/W$, then we can find a D_i which contains O. The frequency with which O intersects the slice S, or equivalently the slice $S_i \subset D_i$, is connected, via the generic mapping degree of $\phi_{i,t} : \theta_i^{-1}(t) \to \chi_i^{-1}(\bar{t})$ (cf. 4.6 Corollary 2), with the frequency with which the preimage $\phi_i^{-1}(S_i)$ intersects that orbit in $G \times^{P_i} A_i$ which covers O. The calculation of the latter will finally enable us to calculate the former.

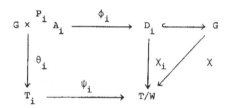

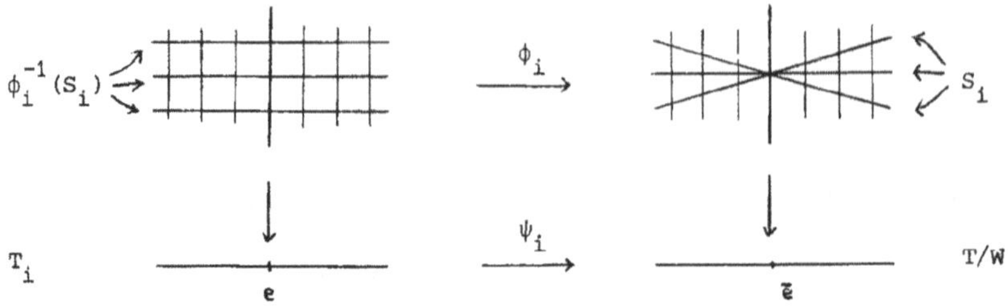

First we determine the "generic" mapping degree of the morphism $\phi_{i,t} : \theta_i^{-1}(t) \to \chi_i^{-1}(\bar{t})$ for $t \in T_i$, i.e. the degree of the étale cover from the dense orbit of $\theta_i^{-1}(t)$ onto its image in $\chi_i^{-1}(\bar{t})$ induced by $\phi_{i,t}$.

Therefore, let $t \in T_i$, and let $\Delta(t) = \Delta_1 \cup \ldots \cup \Delta_m$ be the decomposition of the system of simple roots of T in $B \cap Z(t) \subset Z(t)$ into connected components corresponding to the simple normal subgroup G_j, $j=1,\ldots,m$, of $Z(t)$, which mutually centralize each other. Let $P_i(t) = P_i \cap Z(t)$ and $U_i(t) = U_i \cap Z(t)$. We now define

$$n(t;i,j) = \begin{cases} 0 & \text{if } \alpha_i \notin \Delta_j, \\ 1 & \text{if } \alpha_i \in \Delta_j \text{ and } \Delta_j \text{ homogeneous} \\ & \text{or } \alpha_i \text{ is a short root in } \Delta_j, \\ 2(3) & \text{if } \alpha_i \in \Delta_j \text{ and } \Delta_j \text{ inhomogeneous} \\ & \text{of type } B_{r_j}, C_{r_j}, F_4, (G_2), \\ & \text{and } \alpha_i \text{ is a long root in } \Delta_j. \end{cases}$$

(Note that α_i lies in $\Delta(t)$; $t \in T_i$.) When $t = e$ we have $\Delta(t) = \Delta_1 = \Delta$, so set $n_i = n(e;i,1)$. Then:

$$n_i = \begin{cases} 1 & \text{if } \Delta \text{ homogeneous or } \alpha_i \text{ is a short root,} \\ 2(3) & \text{if } \Delta \text{ inhomogeneous of type } B_r, C_r, F_4, \\ & (G_2) \text{ and } \alpha_i \text{ is a long root.} \end{cases}$$

Lemma 1: Let $u \in Z(t)$ be a subregular unipotent element of type Δ_j. Then u is contained in exactly $n(t;i,j)$ conjugates of $U_i(t)$ under $Z(t)$.

Proof: The element u decomposes into a product $u = u_1 \cdot \ldots \cdot u_m$ of unipotent elements $u_k \in G_k$, where u_k is regular resp. subregular in G_k for $k \neq j$ resp. $k = j$ (cf. 5.4).

Next let $\alpha_i \in \Delta_k$, $k \neq j$. Then u_k is regular in G_k, but $U_i(t) \cap G_k = U_i \cap G_k$ doesn't contain the one-parameter group U_{α_i}. On the basis of the charaterization of regular unipotent elements in G_k (cf. 3.8), u_k will not be contained in any G_k-conjugate of $U_i \cap G_k$ and so u will not be contained in any $Z(t)$-conjugate of $U_i(t)$.

Now let $\alpha_i \in \Delta_j$. Then $P_i \cap G_k$, $k \neq j$, is a Borel subgroup of G_k, and u_k is contained in exactly one G_k-conjugate of $U_i \cap G_k$ (3.8). The group $P_j \cap G_j$ is minimally proper parabolic (i.e. no Borel subgroup), and the element u_j which is subregular in G_j lies in exactly as many G_j-conjugates of $U_i \cap G_j$ as there are lines of type α_i in the Dynkin curve $\mathbb{B}_{u_j} \subset G_j / B \cap G_j$ ([St2] 3.10, Th. 2). That number is just $n(t;i,j)$ (cf. loc. cit. or 6.3).

Proposition 1: Let $t \in T_i$ and $\Delta(t) = \Delta_1 \cup \ldots \cup \Delta_m$ be the Dynkin diagram of $Z(t)$ as given above. Let the root α_i lie in Δ_j. Then $\phi_{i,t}$ has generic degree $n(t;i,j)$.

Proof: If α_i lies in Δ_j, then the dense $P_i(t)$-orbit in $U_i(t)$ consists of just the subregular unipotent elements of type Δ_j in $Z(t)$ (cf. the proof of Lemma 1 and 4.6 Corollary 1). The statement now follows from Lemma 1 and 4.6 Theorem 1 and Corollary 2.

Now let $S_i' = \phi_i^{-1}(S_i)$ be the preimage of S_i in $G \times^{P_i} A_i$.

Lemma 2: The restriction $\phi_i' : S_i' \to S_i$ of the morphism ϕ_i to S_i' is finite.

<u>Moreover S_i' meets only orbits of maximal dimension in $G \times^{P_i} A_i$, and it meets those orbits as a transverse slice.</u>

<u>Proof</u>: Let $s \in S_i$ with $\chi_i(s) = \bar{t} \in T/W$. The preimage $\phi_i^{-1}(O(s))$ of the subregular G-orbit $O(s)$ of s in D_i under the G-equivariant morphism ϕ_i consists of G-orbits of dimension $\geq \dim O(s)$ in $G \times^{P_i} A_i$. Therefore, by 4.6 Corollary 2, it consists of orbits of maximal dimension ($= \dim O(s)$). Since the points of the fiber $\phi_i^{-1}(s)$ thus lie in the dense orbits of some (not necessarily all) of the finitely many fibers $\theta_i^{-1}(t)$, $t \in \psi_i^{-1}(\bar{t})$, the finiteness of the fiber $\phi_i^{-1}(s)$ follows from the finiteness of ϕ_i on the orbits of maximal dimension. Being the restriction over S_i of the proper morphism ϕ_i, the morphism $\phi_i' : S_i' \to S_i$ is likewise proper. Thus by EGA IV 18.12.4, ϕ_i' is finite. It follows from 5.1 Lemma 2 that the morphism $G \times S_i' \to G \times^{P_i} A_i$, $(g,s') \mapsto g \cdot s'$ is smooth. Because of $\dim S_i' = \dim S_i$ and $\dim \phi_i^{-1}(O(s)) = \dim O(s)$, the variety S_i' is a transverse slice to the orbits meeting S_i'.

<u>Lemma 3</u>: <u>The restriction $\theta_i' : S_i' \to T_i$ of the morphism θ_i to S_i' is finite and étale of degree n_i over the neighborhood $\psi_i^{-1}(N)$ of e in T_i.</u>

<u>Proof</u>: The smoothness of θ_i' is proved along the same lines as in the proof of 5.3 Corollary. Since the restriction $\sigma_i : S_i \to T/W$ of σ to the closed subset S_i of the critical locus $\bigcup_{i=1}^{r} S_i$ of σ is finite over the neighborhood N of $\bar{e}$ in T/W (cf. Assumptions), it follows from Lemma 2 that $\psi_i \circ \theta_i' = \sigma_i \circ \phi_i' : S_i' \to T/W$ over N, and so $\theta_i' : S_i' \to T_i$ over $\psi_i^{-1}(N)$, are both finite. Therefore, θ_i' is étale over $\psi_i^{-1}(N)$ of degree $\#\theta_i'^{-1}(e)$. Moreover, $\theta_i'^{-1}(e) = \phi_i^{-1}(x)$ since $\sigma_i^{-1}(\bar{e}) = \{x\}$. By Proposition 1, $\phi_i^{-1}(x)$ consists of just n_i points.

We now come to the last step of the calculations. Let $t \in T$ such that the image $\bar{t}$ of t in T/W lies in N. If the fiber $\chi^{-1}(\bar{t})$ contains subregular orbits, then t is irregular, and any such G-orbit corresponds by the G-isomorphism $G \times^{Z(t)} V(t) \to \chi^{-1}(\bar{t})$ (cf. 3.10) to a subregular unipotent $Z(t)$-orbit in $V(t)$

of a certain type Δ_j , $j \in \{1,\dots,m\}$. By the choice of an appropriate Borel subgroup $B \supset T$ of G we can assume there exists a simple root α_i in the system Δ of simple roots determined by B which lies in the component Δ_j of $\Delta(t)$ (where $\Delta(t)$ is determined by $B(t) = B \cap Z(t))$.

Proposition 2: Suppose the subregular G-orbit O in $\chi^{-1}(\bar{t})$ corresponds to the subregular unipotent $Z(t)$-orbit of type Δ_j , and that there also exists a simple root α_i of Δ which lies in Δ_j . Then O meets the transverse slice S in exactly $n_i/n(t;i,j)$ points.

Proof: Because $\alpha_i \in \Delta_j \subset \Delta(t)$, the element t lies in T_i and O is the image of the dense orbit of the fiber $\theta_i^{-1}(t)$ under $\phi_{i,t}$. If O meets the slice S_i (or S) in n points, then S_i' meets the dense orbit of $\theta_i^{-1}(t)$ in altogether $n(t;i,j)\cdot n$ points (Proposition 1). By Lemmas 2 and 3, the latter intersection points are just the elements of the fiber $\theta_i'^{-1}(t)$ whose cardinality is n_i . Therefore, we have $n = n_i/n(t;i,j)$ (here $n(t;i,j) \neq 0$ as $\alpha_i \in \Delta_j$).

From the definitions we get:

$$n_i/n(t;i,j) = \begin{cases} 2(3) & \text{if } \Delta \text{ is inhomogeneous} \\ & \text{of type } B_r, C_r, F_4, (G_2), \\ & \alpha_i \text{ is a long root in } \Delta \\ & \text{and } \Delta_j, \alpha_i \in \Delta_j, \text{ is homogeneous,} \\ 1 & \text{else.} \end{cases}$$

Proposition 3: Let $C(x)$ be a subgroup of the centralizer $Z_G(x)$ of the subregular unipotent element x which stabilizes the slice S and acts transitively on lines of equal type in the Dynkin curve $\mathbb{B}_x$. Then $C(x)$ also acts transitively on the intersection points of S with a subregular G-orbit of the fiber $\chi^{-1}(\bar{t})$ for all $\bar{t}$ of a neighborhood of $\bar{e}$ in T/W .

<u>Proof:</u> Let the subregular orbit O in $\chi^{-1}(\bar{t})$ lie in the image D_i of the G-morphism ϕ_i. Then along with D_i and S, the group $C(x)$ will also stabilize S_i and $S_i' = \phi_i^{-1}(S_i)$. Moreover, $C(x)$ operates as a group of covering transformations of the êtale (over $\psi_i^{-1}(N)$) covering map $\theta_i' : S_i' \to T_i$. Over a connected neighborhood of e in $\psi_i^{-1}(N)$, this action is determined by the action on the fiber $\theta_i'^{-1}(e) = \phi_i^{-1}(x)$. We may identify G/P_i with the set of lines of type α_i in G/B. After the projection $\phi_i^{-1}(x) \to G \times^P A_i \to G/P_i$ the set $\phi_i^{-1}(x)$ corresponds to the lines of type α_i contained in the Dynkin curve $\mathbb{B}_x$ (cf. Proof of Lemma 1), on which $C(x)$ acts transitively by assumption. Therefore, $C(x)$ acts transitively on the fiber $\theta_i'^{-1}(\bar{e})$, on its neighboring fibers and, by the G-equivariance of ϕ_i', on the intersection points of S with O.

<u>Remark:</u> By means of the explanation given in 4.7, all the results of this section can be translated into the analogous situation for the Lie algebra $\underline{g}$ of G (char(k) very good).

IV Deformations of Simple Singularities

Let k be an algebraically closed commutative field. For technical reasons, $\mathrm{char}(k)$ will be subjected to various restrictions in the individual sections. To study the deformations of simple singularities which arise from transverse slices to the subregular unipotent (resp. nilpotent) orbit of a simple group (resp. its Lie algebra), we next require some information about nilpotent elements.

7. Nilpotent Elements in Simple Lie Algebras.

7.1. Representation Theory for the Lie Algebra sl_2.

Let sl_2 be the Lie algebra of the group SL_2, i.e. the Lie algebra of the 2×2 matrices over k with trace 0. Then there exists a standard basis $sl_2 = kX + kH + kY$ with the elements

$$X = \begin{pmatrix} 0 & 1 \\ 0 & 0 \end{pmatrix}, \quad H = \begin{pmatrix} 1 & 0 \\ 0 & -1 \end{pmatrix}, \quad Y = \begin{pmatrix} 0 & 0 \\ 1 & 0 \end{pmatrix}$$

which satisfy the following commutation relations:

$$[H,X] = 2X \ , \ [H,Y] = -2Y \ , \ [X,Y] = H \ .$$

<u>Definition</u>: A representation $\rho : sl_2 \to gl(V)$ of sl_2 on a finite-dimensional vectorspace V is called <u>good</u>, when either $\mathrm{char}(k) = 0$ or $\mathrm{char}(k) = p > 0$ and $\rho(X)^{p-1} = \rho(Y)^{p-1} = 0$.

Let the d-dimensional vector space V_d be spanned by the basis elements $v_1,\ldots,v_d$. If $\mathrm{char}(k) = p > 0$, let $d < p$. A good, d-dimensional, irreducible representation of sl_2 is defined by:

$$\rho_d(X)v_1 = \rho_d(Y)v_d = 0$$

$$\rho_d(X)v_{i+1} = \overline{i(d-i)}\,v_i \qquad \text{for} \quad 1 \le i \le d-1$$

$$\rho_d(Y)v_i = v_{i+1} \qquad \text{for} \quad 1 \leq i \leq d-1$$

$$\rho_d(H)v_i = \overline{(d-2i+1)}v_i$$

(where for $n \in \mathbb{Z}$, $\overline{n}$ is the residue class modulo p).

Theorem: Every good representation ρ of sl_2 is completely reducible and decomposes into a direct sum of irreducible good represenations of type ρ_d , $d \in \mathbb{N}^+$, ($d < p$ when $\text{char}(k) = p > 0$). In particular, for every natural number $d \in \mathbb{N}^+$, with $d < p$ if $\text{char}(k) = p > 0$, there exists exactly one irreducible representation of dimension d .

A proof of this fact is found in Jacobson [J] or, for $\text{char}(k) = 0$, in LIE VIII § 1.

On the vector space V_d of the good representation ρ_d we can define an action of the multiplicative group G_m by linearly extending the rule

$$(t,v_i) \mapsto t^{d-2i+1} . v_i \ , \ t \in G_m \ , \ 1 \leq i \leq d .$$

If $\rho : sl_2 \to gl(V)$ is any good representation, then we obtain a uniquely defined G_m-action $\lambda : G_m \to GL(V)$ on V by letting G_m operate on each irreducible component of type V_d in the way above. For the differential $D_e\lambda : \text{Lie}(G_m) \to gl(V)$ we then have $D_e(\lambda)(1) = \rho(H)$ (identifying $\text{Lie}(G_m)$ with k).

An element $v \in V$ of a good representation is called an eigenvector of λ or of H of integral weight $n \in \mathbb{Z}$ when

$$\lambda(t)v = t^n v \ , \quad \text{for all} \quad t \in G_m .$$

For $\text{char}(k) = 0$ this condition is equivalent to the infinitesimal condition

$$\rho(H)v = nv .$$

In the following we will understand weights of H as always being integral weights in the above sense. For any $n \in \mathbb{Z}$ let $V(n) = \{v \in V | \lambda(t)v = t^n v\}$ be the <u>eigenspace of weight</u> n.

For later applications we make some simple observations.

Let $V = V_d$ be the vector space of the good representation ρ_d. Then V decomposes into the direct sum of d one-dimensional eigenspace $V(n)$, $n = d-1, d-3, \ldots, -(d-3), -(d-1)$, and the nilpotent endomorphism $\rho_d(X)$ (resp. $\rho_d(Y)$) induces isomorphisms $X : V(n) \xrightarrow{\sim} V(n+2)$ for $n < d-1$ (resp. $Y : V(n) \xrightarrow{\sim} V(n-2)$ for $n > -(d-1)$) and annihilates $V(d-1)$ (resp. $V(-(d-1))$).

If d is even (resp. odd), then $\dim V_d(1) = 1$ (resp. $\dim V_d(0) = 1$). The number of irreducible summands in the decomposition of a good representation V is therefore $\dim V(0) + \dim V(1)$.

7.2. The Jacobson-Morozov Lemma.

Let $\underline{g}$ be the Lie algebra of the simple group G (of type Δ) and let $x \in N(\underline{g})$ be a nilpotent element in $\underline{g}$.

<u>Definition</u>: A triple (x,h,y) of elements of $\underline{g}$ is called an <u>sl_2-triplet</u> for (the nilpotent element) x, when there is a homomorphism of Lie algebras $\rho : sl_2 \to \underline{g}$ with $\rho(X) = x$, $\rho(H) = h$, $\rho(Y) = y$.

<u>Remark</u> : For $\mathrm{char}(k) \neq 2$, the Lie algebra sl_2 is simple. Therefore, in this case the homomorphism $\rho : sl_2 \to \underline{g}$ is either injective or zero.

The <u>Coxeter number</u> $\mathrm{Cox}(G)$ <u>or</u> $\mathrm{Cox}(\Delta)$ <u>of</u> G <u>or</u> Δ is $m + 1$, where m is the height of the largest root in the root system belonging to G. We have the following values for $\mathrm{Cox}(G)$ (cf. LIE IV, V, VI, Planches):

type of G	A_r	B_r	C_r	D_r	E_6	E_7	E_8	F_4	G_2
$\mathrm{Cox}(G)$	$r+1$	$2r$	$2r$	$2r-2$	12	18	30	12	6

Theorem (Jacobson-Morozov, cf. [S-S] III. 4.3): Let char(k) = 0 or > 4 Cox(G) - 2. For every nilpotent element $x \in \underline{g}$ there exists an sl_2-triplet, and the composition of the associated representation $\rho : sl_2 \to \underline{g}$ with the adjoint representation $ad : \underline{g} \to gl(\underline{g})$ is good.

7.3. The Classification of Nilpotent Elements.

Let G be a simple adjoint group; and let char(k) = 0 or > 4 Cox(G) - 2 . In particular, char(k) will be very good (cf. 3.13). Let x be a nilpotent element of the Lie algebra $\underline{g}$ of G . By the theorem in 7.2 there exists an sl_2-triplet in $\underline{g}$ and $\underline{g}$ decomposes relative to ad(h) into a direct sum $\underline{g} = \bigoplus_{i \in \mathbb{Z}} \underline{g}(i)$ of eigenspaces of integral weights (cf. 7.1). Moreover, we will have $[\underline{g}(i), \underline{g}(j)] \subset \underline{g}(i+j)$. The G_m-action $\lambda : G_m \to GL(\underline{g})$ defined as in 7.1 for the good sl_2-representation on $\underline{g}$ will then commute with the Lie bracket of $\underline{g}$. Therefore, λ factors through a one-parameter group $G_m \to Aut^o(\underline{g})$, which we will also denote by λ , into the inner automorphisms G of $\underline{g}$.

Such a one-parameter group λ is said to be adapted to x . We now have ([S-S] III 4.):

Lemma 1: Let x and x' be nilpotent elements in $\underline{g}$. Then the following properties are equivalent:

(i) x and x' are conjugate under G .

(ii) Any sl_2-triplet of x is conjugate to any sl_2-triplet of x' under G .

(iii) Any one-parameter group adapted to x is conjugate to any such group adapted to x' under G .

(Two triplets (x,h,y) , (x',h',y') are called conjugate when there is a $g \in G$ with gx = x' , gh = h' , gy = y' .)

The one-parameter group λ defines a parabolic subgroup P in G with Lie algebra

$Lie(P) = \bigoplus_{i \geq 0} \underline{g}(i)$. If T is a maximal torus of P , and so of G , which contains $\lambda(G_m)$, and if Δ is the system of simple roots of T in G determined by a Borel subgroup $T \subset B \subset P$, then it can be shown that the numbers $\alpha(\lambda)$, $\alpha \in \Delta$, can only have the values 0,1 or 2 (cf. [S-S] III 4., or for char(k) = 0 , LIE VIII § 11). We will order the vertices of the Dynkin diagram of G by means of an "épinglage" (cf. LIE VIII § 4, n° 1).

Definition: The ordered Dynkin diagram of G valuated by the integers $\alpha(\lambda)$, $\alpha \in \Delta$, is called the Dynkin diagram $\Delta(x)$ of x with respect to λ , T and B .

The following result goes back to Dynkin in the case char(k) = 0 (cf. [Dy] Chap. III § 8, [S-S] III. 4.24, 4.25).

Theorem: Let $x \in \underline{g}$ be nilpotent. Then the Dynkin diagram $\Delta(x)$ is independent of the choice of λ , T and B . Two nilpotent elements in $\underline{g}$ are conjugate exactly when their Dynkin diagrams coincide.

Corollary 1: If r is the rank of G , then there are at most 3^r different conjugate classes of nilpotent elements in $\underline{g}$.

In fact, not all possible 3^r diagrams will appear as Dynkin diagrams of nilpotent elements (cf. [Dy] pp. 176 - 185).

Let (x,h,y) be an sl_2-triplet for the nilpotent element $x \in \underline{g}$. With that, $\underline{g}$ becomes an sl_2-module which decomposes into a direct sum of irreducible submodules $\underline{g} = \bigoplus_{j=1}^{s} E_j$ (7.2). Each of these irreducible submodules E_j will contain a unique one-dimensional eigenspace $E_j(n_j)$ of h (with highest weight $n_j = \dim E_j - 1$) which is annihilated by x (cf. 7.1). Therefore, the dimension of the centralizer of x in $\underline{g}$ is equal to the number s of irreducible components E_j , which in turn is equal to $\dim \underline{g}(0) + \dim \underline{g}(1)$. From the decomposition $\underline{g} = \underline{h} \oplus \bigoplus_{\alpha \in \Sigma} \underline{g}_\alpha$, where $\underline{h} = Lie(T)$, and with $\lambda : G_m \to T$ adapted to x it follows that $\dim Z_G(x) = \dim \underline{z}_{\underline{g}}(x) = r + \#\{\alpha \in \Sigma \mid \alpha(\lambda) \in \{0,1\}\}$. The values $\beta(\lambda)$, $\beta \in \Sigma$, are determined by the values $\alpha(\lambda)$, $\alpha \in \Delta$. Therefore, $\dim Z_G(x)$ can be calculated from $\Delta(x)$.

Corollary 2: Let $x \in \underline{g}$ be a regular nilpotent element. Then the valuations of the simple roots of the Dynkin diagram $\Delta(x)$ are all equal to 2 .

Proof: The equation $r = \dim Z_G(x) = r + \#\{\alpha \in \Sigma \mid \alpha(\lambda) \in \{0,1\}\}$ excludes the values 0 and 1 for $\alpha(\lambda)$, $\alpha \in \Delta$.

Lemma 2: Let G be simple and let $x \in \underline{g}$ be a subregular element, i.e. $\dim Z_G(x) = r+2$. Then x has the following Dynkin diagram $\Delta(x)$ according to the type of G :

A_{2n+1}	$n \geq 0$	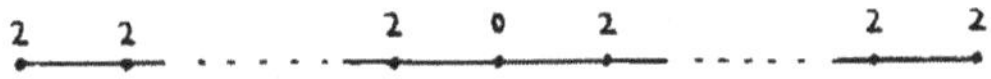
A_{2n}	$n \geq 1$	
B_r	$r \geq 2$	
C_r	$r \geq 3$	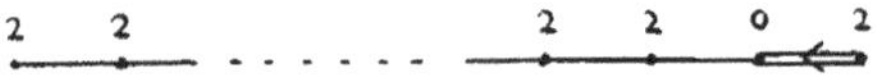
D_r	$r \geq 4$	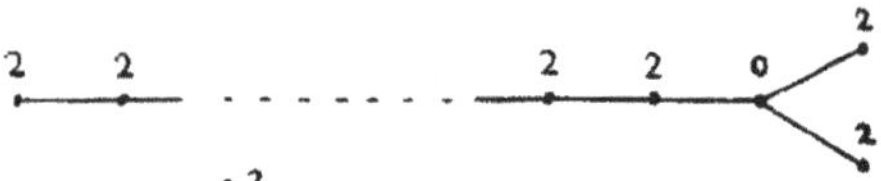
E_6		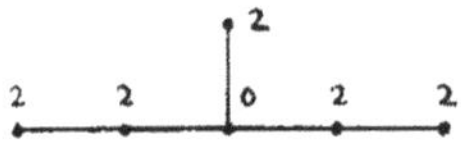
E_7		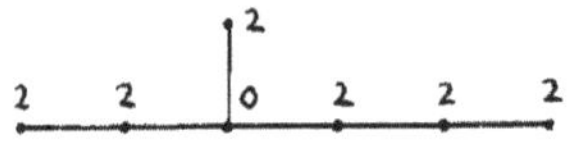
E_8		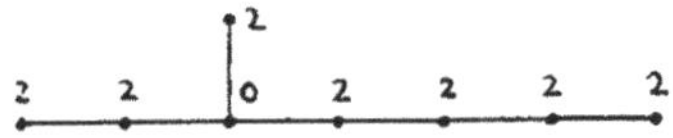
F_4		
G_2		

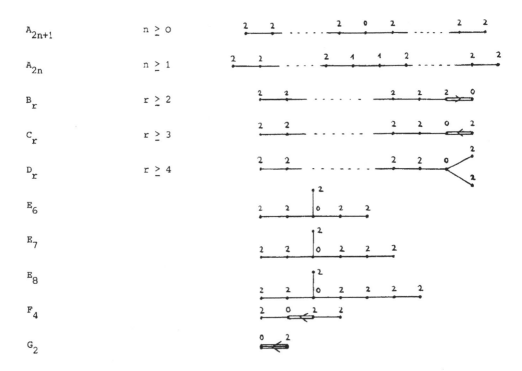

Proof: For the classical types A_r , B_r , C_r , D_r the Dynkin diagram of x may be derived from the Jordan normal form of x in the natural representation of $\underline{g}$

(cf. [S-S] IV and [He] 4.5.6). For the exceptional types $E_6, E_7, \ldots, G_2$, Dynkin's tables ([Dy] pp. 176-185) of all possible diagrams are at our disposal. The statement follows either from an easy application of the formula for the centralizer dimension or from the uniqueness of the subregular class (5.4 and the G-isomorphism between $V(G)$ and $N(\underline{g})$) combined with Table 21 in [Dy].

7.4. A Special Transverse Slice.

We consider the same situation as we did in the beginning of 7.3 and use the same notations and assumtions.

Let (x,h,y) be an sl_2-triplet for the nilpotent element $x \in \underline{g}$. In this section we construct a technically useful transverse slice S to the G-orbit of x in $\underline{g}$.

The calculation of the differential of the orbit map $G \to \underline{g}$, $g \mapsto g \cdot x = Ad(g)x$, shows that $x + (ad\ x)(\underline{g})$ is the affine tangent space to the orbit of x at the point x (cf. [Bo] I. 3.9(2)). Let $\underline{g} = \bigoplus_{j=1}^{s} E_j$ be a direct decomposition of $\underline{g}$ into irreducible sl_2-modules of dimension $\dim E_j = n_j + 1$. Any one of the one-dimensional subspaces $E_j(m)$, $m \geq -n_j + 2$, lies in the image of the endomorphism $ad\ x$, namely $E_j(m) = (ad\ x)\, E_j(m-2)$ (cf. 7.1). Therefore, $E_j(-n_j)$ is a complement to $(ad\ x)\, E_j$ in E_j and $\underline{z}_{\underline{g}}(y) = \bigoplus_{j=1}^{s} E_j(-n_j)$ is a complement to $(ad\ x)\,\underline{g}$ in $\underline{g}$. Then $S = x + \underline{z}_{\underline{g}}(y)$ is a subvariety of $\underline{g}$ that will be a transverse slice to the G-orbit of x in the sense of 5.3 for some neighborhood of the point x at least (later we will see that S is already transverse everywhere).

Definition: A polynomial $f \neq 0$ of $k[X_1,\ldots,X_s]$ is called quasihomogeneous of type $(d; w_1,\ldots,w_s)$ for integers $d, w_1,\ldots,w_s$ when all the monomials $a_{i_1,\ldots,i_s} X_1^{i_1} \ldots X_s^{i_s}$ with $a_{i_1,\ldots,i_s} \neq 0$ which appear in f satisfy $\sum_{j=1}^{s} i_j w_j = d$. A morphism $F : \mathbb{A}^s = \operatorname{Spec} k[X_1,\ldots,X_s] \to \mathbb{A}^r = \operatorname{Spec} k[Y_1,\ldots,Y_r]$ is called a G_m-morphism or quasihomogeneous of type $(d_1,\ldots,d_r\,;\, w_1,\ldots,w_s)$ when the components $f_i = Y_i \circ F \in k[X_1,\ldots,X_s]$ are quasihomogeneous of type $(d_i; w_1,\ldots,w_s)$. The d_i are called the degrees, and the w_j the weights, of F.

Defining a G_m-action on $\mathbb{A}^s$ and $\mathbb{A}^r$ by $t\cdot(x_1,\dots,x_s) = (t^{w_1}x_1,\dots,t^{w_s}x_s)$ and $t\cdot(y_1,\dots,y_r) = (t^{d_1}y_1,\dots,t^{d_r}y_r)$, $t \in G_m$, we see that the quasihomogeneity of $F : \mathbb{A}^s \to \mathbb{A}^r$ is equivalent to the G_m-equivariance of F.

We again consider $S = x + \underline{z}_{\underline{g}}(y)$. Let $e_j \neq 0$, $j = 1,\dots,s$, be elements from $E_j(-n_j)$ and let X_j be their dual functionals, $X_j(e_i) = \delta_{ij}$. By the choice of this basis for $\underline{z}_{\underline{g}}(y)$ and by the isomorphism $\underline{z}_{\underline{g}}(y) \to S$, $z \mapsto x + z$, we can identify the regular functions $k[S]$ on S with $k[X_1,\dots,X_s]$ and S with $\mathbb{A}^s$. Similarly, by the choice of homogeneous generators $\gamma_1,\dots,\gamma_r$ of $k[\underline{g}]^G = k[\underline{h}]^W$ (cf. 3.12) we can identify $k[\underline{h}/W]$ with $k[Y_1,\dots,Y_r]$ and $\underline{h}/W$ with $\mathbb{A}^r$. Respecting these identifications and notations we have:

<u>Proposition 1:</u> <u>The restriction</u> $\delta : S \to \underline{h}/W$ <u>of the adjoint quotient</u> $\gamma : \underline{g} \to \underline{h}/W$ <u>of</u> $\underline{g}$ to $S = x + \underline{z}_{\underline{g}}(y)$ <u>is quasihomogeneous of type</u> $(d_1,\dots,d_r;w_1,\dots,w_s)$, <u>where</u> $d_i = 2m_i + 2$ <u>is the homogeneous degree of</u> γ_i <u>multiplied by two, and where</u> $w_j = n_j + 2$.

<u>Proof:</u> Let $\lambda : G_m \to G$ be the one-parameter group adapted to the nilpotent element x (cf. 7.3) which operates on $\underline{g}(m)$ with weight m. The usual scalar action of G_m on the vector space $\underline{g}$ defines an additional action $G_m \times \underline{g} \to \underline{g}$, $(t,v) \mapsto \sigma(t)v = tv$. Because of the G-invariance resp. the homogeneity of the polynomials γ_i, we have

$$\gamma_i(\lambda(t)v) = \gamma_i(v) \quad \text{resp.} \quad \gamma_i(\sigma(t)v) = t^{m_i+1}\gamma_i(v) \qquad i=1,\dots,r .$$

Let $v = x + \sum_{i=1}^{s} x_j e_j$ be an element of $S = x + \underline{z}_{\underline{g}}(y)$. Then

$$\lambda(t)v = t^2 x + \sum_{i=1}^{s} t^{-n_j} x_j e_j$$

and

$$\sigma(t^{-2})\lambda(t)v = x + \sum_{j=1}^{s} t^{-(2+n_j)} x_j e_j \in S .$$

Since σ and λ commute, the map $t \mapsto \rho(t) = \sigma(t^2)\lambda(t^{-1})$ defines a G_m-action $G_m \to GL(\underline{g})$ that fixes x, stabilizes S, and by the foregoing identification of S with $\mathbb{A}^s$ has the form

$$t \cdot (x_1,\dots,x_s) = (t^{n_1+2} \cdot x_1,\dots,t^{n_s+2} \cdot x_s) \quad \text{on} \quad S .$$

However, the polynomials γ_i, $i=1,\dots,r$, satisfy

$$\gamma_i(\rho(t)v) = \gamma_i(\sigma(t^2)v) = t^{2(m_i+1)}\gamma_i(v) = t^{d_i}\gamma_i(v) .$$

So the morphism $\delta = \gamma|_S : S \to \underline{h}/W$ is quasihomogeneous of type $(d_1,\dots,d_r;n_1+2,\dots,n_s+2)$.

Corollary 1: The morphism $\delta : S \to \underline{h}/W$ is faithfully flat and $G \times S \to \underline{g}$, $(g,s) \mapsto g \cdot s$, is smooth at all points. If x is subregular then $\delta^{-1}(\overline{0})$ has exactly one singular point (at x).

Proof: The flatness of δ will follow from the smoothness statement similarly as in 5.2. As a result, $\delta(S)$ will be an open neighborhood of $\overline{0} \in \underline{h}/W$ which will be stable under G_m since δ is G_m-equivariant. Since the G_m-weights on $\underline{h}/W$ are positive this neighborhood must be equal to $\underline{h}/W$. Hence δ is surjective and thus faithfully flat. Let G_m act on $\underline{g}$ by $t \mapsto \sigma(t^2)$ and on $G \times S$ by $(t,(g,s)) \mapsto (g\lambda(t),\rho(t)s)$ (notation as in the proof of Proposition 1). Then the morphism $\mu : G \times S \to \underline{g}$, $(g,s) \mapsto g \cdot s$, will be equivariant with respect ot these actions. Therefore, the set of smooth points of μ is stable under G_m as well as under the G-action by left translation on the left factor of $G \times S$. Since μ is smooth in a neighborhood of $G \times \{x\}$ and since the weights of the ρ-action on S are strictly positive, it follows that μ is smooth on all of $G \times S$.

From the fact that $\delta^{-1}(\overline{0})$ has only isolated singularities for subregular x (cf. 5.5) and is stable under the ρ-action of G_m on S, the last statement follows, again because the weights are strictly positive.

Remark: The statement in the proposition goes back to a more general statement about differential operators by Harish-Chandra ([H-C] Lemma 30), and was derived more simply by Varadarajan ([V]), however only for the analytical case over $\mathbb{C}$, to use in studying γ at the regular elements of $\underline{g}$.

Proposition 2: Let x be a nilpotent element of the simple Lie algebra $\underline{g}$, and let $\delta : S \to h/W$ be as in Proposition 1. If x is regular, then the weights of δ equal the degrees $w_i = d_i$, $i=1,\dots,r$. If x is subregular, then the weights $w_1,\dots,w_{r+2}$ are given in the table below, which also lists the degrees $d_1,\dots,d_r$:

	d_1	d_2	d_3	....		d_{r-2}	d_{r-1}	d_r	w_r	w_{r+1}	w_{r+2}
A_r	4	6	8	....		2r-2	2r	2r+2	2	r+1	r+1
B_r	4	8	12	....		4r-8	4r-4	4r	2	2r	2r
C_r	4	8	12	....		4r-8	4r-4	4r	4	2r-2	2r
D_r	4	8	12	....		4r-8	2r	4r-4	4	2r-4	2r-2
E_6	4	10	12			16	18	24	6	8	12
E_7	4	12	16		20	24	28	36	8	12	18
E_8	4	16	24	28	36	40	48	60	12	20	30
F_4	4	12					16	24	6	8	12
G_2	4							12	4	4	6

Furthermore $w_i = d_i$ for $i = 1,\dots,r - 1$.

Proof: The values $d_i = 2m_i + 2$, $i=1,\dots,r$ may be obtained, for example, from the tables of Bourbaki (LIE IV, V, VI, Planches) for the exponents m_i . To calculate w_j , $j=1,\dots,r+2$, consider the sl_2-triplet (x,h,y) for x . From the Dynkin diagram $\Delta(x)$ of x we obtain the values $\alpha(\lambda) = \alpha(h)$ for all roots α in a basis Δ of the root system of T in $\underline{g}$ (here T is a maximal torus of G with Lie $T = h$, and $\lambda : G_m \to T \quad G$ is adapted to x) (cf. 7.3). Because of the decomposition $\underline{g} = \underline{h} \oplus \bigoplus_{\alpha \in \Sigma} \underline{g}_\alpha$ and the additivity $(\alpha+\beta)(h) = \alpha(h) + \beta(h)$, $\alpha, \beta \in \Sigma$, we may then calculate all of the eigenvalues of $ad(h)$ on $\underline{g}$, i.e. the dimensions of the eigenspaces $\underline{g}(m)$, $m \in \mathbb{Z}$. The multiplicity with which an irreducible module of highest weight $n \in \mathbb{N}$ in $\underline{g}$ appears is now given by $\dim \underline{g}(n) - \dim \underline{g}(n+2)$. This follows from the description of the structure of a good sl_2-module given in 7.1.

For the exceptional types E_6, E_7, E_8, F_4, G_2 on may find the calculations of the highest weights $n_1, \dots, n_{r+2}$ done in table 21 of [Dy] pp. 186-187. The weights of δ will then be $w_j = n_j + 2$. The calculations for the classical types are left to the reader.

Example: To illustrate the calculation method described in the proof above we apply it to the subregular class in A_3, whose Dynkin diagram has the form

Let $\alpha_1, \alpha_2, \alpha_3$ be the simple roots corresponding (in order) to the vertices of this diagram. All positive roots are then given by

$$\Sigma^+ = \left\{ \begin{array}{ccccc} \alpha_1 & & \alpha_2 & & \alpha_3 \\ & \alpha_1 + \alpha_2 & & \alpha_2 + \alpha_3 & \\ & & \alpha_1 + \alpha_2 + \alpha_3 & & \end{array} \right\}$$

Thus we obtain the following eigenvalues of $ad(h)$ on $\underline{n}^+ = \bigoplus_{\alpha \in \Sigma^+} \underline{g}_\alpha$:

$$\begin{array}{ccccc} 2 & & 0 & & 2 \\ & 2 & & 2 & \\ & & 4 & & \end{array}$$

resp. on $\underline{n}^- = \bigoplus_{\alpha \in \Sigma^-} \underline{g}_\alpha$:

$$\begin{array}{ccccc} -2 & & 0 & & -2 \\ & -2 & & -2 & \\ & & -4 & & \end{array}$$

On $\underline{h}$ we get three times 0 . Hence $\underline{g}$ splits into one sl_2-module of highest weight 4, three of highest weight 2, and one of highest weight 0 , leading to the values 6,4,4,4,2 for the weights of δ . The following corollary is a special case of the more general differential criterion of regularity (cf. 3.8, 3.14).

Corollary 2: Let $x \in \underline{g}$ be a regular nilpotent element. Then $\delta : S \to \underline{h}/W$ is an isomorphism.

Proof: We know that δ is flat (Corollary 1) and that $w_i = d_i$, $i=1,\dots,r$ for the weights and degrees of δ (Proposition 2). The result now follows from the (technical) Lemma 3 , 8.1.

Remark: In the work of Kostant and Varadarajan ([Ko1], [Ko2], [V]) this corollary is the key step in the proof of the differential criterion for regular elements.

7.5. Centralization of Nilpotent Elements.

Again let G be simple, adjoint with Lie algebra $\underline{g}$ and $\mathrm{char}(k) = 0$ or $> 4 \cdot \mathrm{Cox}(G) - 2$. Let (x,h,y) be an sl_2-triplet for the nilpotent element $x \in \underline{g}$, and $\lambda : G_m \to G$ a one-parameter group adapted to x with $\mathrm{Lie}(\lambda(G_m)) = k \cdot h$. Let $\underline{z}$ resp. $\underline{c}$ resp. $\underline{r}$ denote the Lie algebra of the centralizer $Z = Z_G(x)$, resp. $C = Z_G(x) \cap Z_G(h)$ resp. the unipotent radical R of Z . Under the action of h , or equivalently $\lambda(G_m)$, the Lie algebra $\underline{g}$ decomposes into a direct sum of eigenspaces $\underline{g} = \bigoplus_{n \in \mathbb{Z}} \underline{g}(n)$.

Lemma 1: (i) The image $\lambda(G_m)$ normalizes Z. The weights of h in $\underline{z}$ are ≥ 0 and those of h in $\underline{r}$ are > 0.

(ii) The identity component C^o of C is reductive, and Z is the semidirect product $Z = C \ltimes R$ of C and R.

(iii) The component group Z/Z^o of Z is isomorphic to that of C.

For the proof see [S-S] III 4.10 and [El] 5.1 - 5.4.

Definition: Because of property (ii) we call C the reductive centralizer of x (with respect to h).

Lemma 2: The reductive centralizer of x centralizes the sl_2-subalgebra of g generated by (x,h,y).

Proof: Let $c \in C$. Then $(cx,ch,cy) = (x,h,cy)$ is also an sl_2-triplet for x. Since $\mathrm{ad}\, x : \underline{g}(-2) \to \underline{g}(0)$ is injective (cf. 7.1) and $h = [x,y] = [x,cy]$, it follows that $cy = y$ and so the lemma holds.

Remark: We also have $C = Z_G(x) \cap Z_G(y)$ by Lemma 2 and the fact that $[x,y] = h$.

From the description of C above it follows that C stabilizes the transverse slice $S = x + \underline{z}_{\underline{g}}(y)$, and that the morphism $\delta : S \to \underline{h}/W$ (cf. 7.4) is invariant with respect to the C-action on S. Since C commutes with both λ and the scalar action σ of G_m on $\underline{g}$ (linearity of the adjoint representation), we get a $C \times G_m$-action on S, where the action of the factor G_m comes from the action $\rho = \sigma^2\lambda^{-1}$ as in 7.4 Proposition 1.

Now let x be subregular. On the basis of the G-isomorphism between $N(\underline{g})$ and $V(G)$, the resolution of $V(G)$ in 4.1 also gives a resolution of $N(\underline{g})$. Because this resolution is G-equivariant, the group $Z_G(x)$ operates on the exceptional fiber $\mathcal{B}_x$ over x which is a Dynkin curve (cf. 6.3).

Lemma 3: The group $Z_G(x)$ operates transitively on lines of the same type in the Dynkin curve over x.

This lemma is a reformulation of a statement in 4.6 Theorem 1, cf. also [St2] 3.10, Prop. 1(d).

If G has a homogeneous Dynkin diagram Δ, then every line type appears exactly once in the Dynkin curve of type Δ. So the result above is of interest only in case of groups with inhomogeneous Dynkin diagram.

Since, by continuity, the identity component of the centralizer $Z_G(x)$ cannot permute the components of the Dynkin curve, it follows that for groups with inhomogeneous Dynkin diagrams, $Z_G(x)$ is not connected. More precisely, for the reductive centralizer C of x, which is relevant for this question, we have:

Lemma 4: If $x \in \underline{g}$ is a regular nilpotent element, then $C = \{e\}$, in particular $Z_G(x)$ is connected. If x is subregular, then according to type we have:

Δ	$A_r, r>1$	B_r	C_r	D_r	E_6	E_7	E_8	F_4	G_2
C	G_m	$G_m \rtimes \mathbb{Z}/2.\mathbb{Z}$	$\mathbb{Z}/2.\mathbb{Z}$	$\{e\}$	$\{e\}$	$\{e\}$	$\{e\}$	$\mathbb{Z}/2.\mathbb{Z}$	$\mathfrak{S}_3$

(In the semidirect product $G_m \rtimes \mathbb{Z}/2\,\mathbb{Z}$, $\mathbb{Z}/2\,\mathbb{Z}$ acts on G_m in the unique nontrivial fashion $x \mapsto x^{-1}$.)

Proof: For regular x the result is found in [S-S] III 3.7. In the subregular case the statement that $\dim C = 0$ for the exceptional types appears in [El] as well as the calculation that $C = \mathfrak{S}_3$ (symmetric group) for G_2 ([El] § 6). The case F_4, E_6, E_7, E_8 will be done later (8.5) (F_4 is also found in [Sh] Table 7). The statement for A_r follows from [S-S] III 3.22, IV 1.8 (the centralizer in PGL_{r+1} is the image of the centralizer in GL_{r+1}). For the remaining classical cases we use [S-S] IV 2.25. The Jordan blocks of the subregular nilpotent elements in the Lie algebras so_{2r+1} , sp_{2r} , so_{2r} are given by the partitions (2r-1,1,1) , (2r-2,2) , (2r-3,3) ([He] 4.5.6 , or 7.3 Corollary 2 and [S-S] IV 2.33). Using [S-S] IV 2.14, 2.25, we obtain for the subgroup C in O_{2r+1} , Sp_{2r} , O_{2r} correspondingly the groups $O_1 \times O_2$, $O_1 \times O_1$, $O_1 \times O_1$. Here $O_1 \cong \mathbb{Z}/2\mathbb{Z}$ and $O_2 \cong G_m \rtimes \mathbb{Z}/2\mathbb{Z}$. Passing next to the special orthogonal groups (in the case B_r note loc. cit. 2.19) and then to the adjoint groups (i.e. quotient of Sp_{2r} and SO_{2r} by the center $\{+1,-1\}$) we get the stated facts.

Proposition: Let Δ be an inhomogeneous irreducible Dynkin diagram and x a subregular nilpotent element in a Lie algebra $\underline{g}$ of type Δ . Then $C(x)/C(x)^{\circ} = Z(x)/Z(x)^{\circ}$ is isomorphic to $AS(\Delta)$ and the action of $C(x)/C(x)^{\circ}$ on the dual diagram $_h\Delta$ of the Dynkin curve $\mathcal{B}_x$ coincides with the associated action of $AS(\Delta)$ on $_h\Delta$.

Proof: The statements follow from Lemma 3 and 4 and the definitions (cf. 6.2, 6.3).

Remark: The automorphism group of a Dynkin curve considered as a variety can be easily calculated (use [Bo] III 10.8), and for homogeneous Δ of rank r it has dimension $r + 2$. However it is not isomorphic to $Z_G(x)$ since its reductive part contains a torus of rank $r - 1$ (even of rank r for $\Delta = A_r$) in contrast to $Z_G(x)$.

7.6. Outer Centralization of Nilpotent Elements.

We consider the same situation as in 7.5. Besides the centralizer in G of a nilpotent element $x \in \underline{g}$, we will also be interested in the isotropy group of x with respect to the action of all automorphisms of $\underline{g}$ on $\underline{g}$. It is known that there is a split exact sequence

$$1 \to G \to \mathrm{Aut}(\underline{g}) \to \mathrm{Aut}(\Delta) \to 1$$

where $\mathrm{Aut}(\Delta)$ is the (finite) group of automorphisms of the Dynkin diagram Δ of

G (cf. LIE VIII § 5, n° 2.3, [St 0]).

Let $ZA(x) := \{\sigma \in \mathrm{Aut}(\underline{g}) \mid \sigma(x) = x\}$ be the isotropy group of x in $\mathrm{Aut}(\underline{g})$. If (x,h,y) is an sl_2-triplet for x, we define $CA(x) := \{\sigma \in ZA(x) \mid \sigma(h) = h\}$ to be the outer reductive centralizer of x (with respect to h). Obviously $CA(x)$ stabilizes each h-eigenspace in $\underline{g}$.

Lemma 1: The group $ZA(x)$ is the semidirect product $ZA(x) = R(x) \rtimes CA(x)$ of its unipotent radical $R(x)$ with $CA(x)$, and $ZA(x)/ZA(x)^{\circ} \cong CA(x)/CA(x)^{\circ}$. The outer reductive centralizer $CA(x)$ centralizes the sl_2-subalgebra of $\underline{g}$ generated by (x,h,y).

Proof: The proofs of Elkington [El] 5.3, 5.4 as well as the proof of 7.5 Lemma 2 concerning the centralizer in G can be applied verbatim to the situation above.

Similarly as in 7.5 we see that $CA(x)$ stabilizes the transverse slice $S = x + \underline{z}_{\underline{g}}(y)$ and commutes with the G_m-action on S (7.4 Proposition 1).

Lemma 2: If $x \in \underline{g}$ is a regular nilpotent element, then $CA(x)$ is isomorphic to $\mathrm{Aut}(\Delta)$. If $x \in \underline{g}$ is a subregular nilpotent element, then $CA(x)$ has the following form depending on the type of Δ:

Δ	$A_r, r>1$	D_4	$D_r, r>4$	E_6	other types
$CA(x)$	$E(\mathbb{Z}/2\mathbb{Z}, G_m)$	$\mathfrak{S}_3$	$\mathbb{Z}/2\mathbb{Z}$	$\mathbb{Z}/2\mathbb{Z}$	$C(x)$

Here $E(\mathbb{Z}/2\mathbb{Z}, G_m)$ is an extension of $\mathbb{Z}/2\mathbb{Z}$ by G_m with nontrivial action of $\mathbb{Z}/2\mathbb{Z}$ on G_m. For r odd this extension is semidirect, for r even it is the unique nontrivial extension.

Proof: We may assume $\Delta = A_r, r>1$, D_r, E_6 since in the other cases there is nothing to prove. Let $\sigma \in \mathrm{Aut}(\underline{g})$ be an arbitrary automorphism of $\underline{g}$. Then

$(\sigma(x),\sigma(h),\sigma(y))$ is an sl_2-triplet for $\sigma(x)$, which is again regular resp. subregular. By the uniqueness of the regular resp. subregular nilpotent G-orbit in $\underline{g}$ (cf. 7.3) there is a $g \in G$ with $g \circ \sigma(x,h,y) = (x,h,y)$. So $g\sigma$ lies in $CA(x)$. From the surjectivity of $Aut(\underline{g}) \to Aut(\Delta)$ it follows that the composition $CA(x) \to Aut(\underline{g}) \to Aut(\Delta)$ is surjective. For regular x as well as for $\Delta = D_r$ or E_6 and x subregular the kernel $C(x) = CA(x) \cap G$ of this map is trivial (7.5 Lemma 4), and so $CA(x) = Aut(\Delta)$. For $\Delta = A_r, r>1$, and x subregular we have $C(x) = G_m$ and $Aut(\Delta) = \mathbb{Z}/2\mathbb{Z}$. Thus $CA(x)$ is an extension of $\mathbb{Z}/2\mathbb{Z}$ by G_m. Let $s \in CA(x)$ be an element which is not contained in $C(x)$. To prove the remaining statements we have to show that conjugation by s induces the nontrivial automorphism of $C(x) \cong G_m$ and that $s^2 = 1$ resp. $s^2 = -1$ for r odd resp. r even. It is sufficient to verify these properties for an example.

Let $x \in sl_{r+1} = pgl_{r+1}$ be the subregular element given by the matrix

$$\left(\begin{array}{cccc|c} 0 & 1 & & & \\ & \ddots & \ddots & & \\ & & \ddots & 1 & \\ & & & 0 & \\ \hline & & & & 0 \end{array}\right)$$

and let $\sigma \in Aut(sl_{r+1})$ be the outer automorphism $\sigma(m) = {}^t(-m)$. Let

$$g = \left(\begin{array}{cccc|c} & & & \pm 1 & \\ & & \cdot^{\cdot^{\cdot}} & & \\ & +1 & & & \\ -1 & & & & \\ +1 & & & & \\ \hline & & & & 1 \end{array}\right) \in GL_{r+1}$$

and $\phi := (\operatorname{Int} g) \circ \sigma \in Aut(sl_{r+1})$. Then $\phi(x) = x$ and $\phi^2 = 1$ resp. $\phi^2 = -1$ if r is odd resp. if r is even. With respect to a suitable triplet (x,h,y) the Lie algebra $\underline{c}(x)$ of the reductive centralizer is given by

$$\underline{c}(x) = \left\{ m(t) = \left(\begin{array}{cccc|c} t & & & & \\ & \ddots & & & \\ & & \ddots & & \\ & & & t & \\ \hline & & & & -rt \end{array}\right) \;\middle|\; t \in k \right\} .$$

Then ϕ transforms $\underline{c}(x)$ into itself mapping $m(t)$ to $m(-t)$. With that the proof of the lemma is complete.

Let $\mathcal{B}$ denote the variety of all Borel subalgebras of $\underline{g}$, and let $\pi : Y \to N(\underline{g})$ be the resolution of the nilpotent variety, where $Y \subset \mathcal{B} \times N(\underline{g})$ is the subvariety of pairs $(\underline{b},x)$ with $x \in \underline{b}$ and where π is induced by the second projection (cf. 4.1, 4.7). If we let $\mathrm{Aut}(\underline{g})$ act on $\mathcal{B} \times N(\underline{g})$ by $\sigma(\underline{b},x) = (\sigma(\underline{b}),\sigma(x))$ for $\sigma \in \mathrm{Aut}(\underline{g})$, then Y will be $\mathrm{Aut}(\underline{g})$-stable and π will be $\mathrm{Aut}(\underline{g})$-equivariant. Let x be a subregular nilpotent element in $\underline{g}$. Then $ZA(x)$ acts on the fiber $\pi^{-1}(x) = \mathcal{B}_x$ which is a Dynkin curve (cf. 6.3). Let Δ be a system of simple roots corresponding to the choice of a Cartan-algebra $\underline{h}$ and a Borel algebra $\underline{b}$ in $\underline{g}$, and let $\sigma \in ZA(x)$ with image σ_o in $\mathrm{Aut}(\Delta)$. If $\underline{p}$ is a parabolic subalgebra of the form $\underline{p} = \underline{b} \oplus \underline{g}_{\alpha}$, where $\underline{g}_{\alpha}$ is the onedimensional root space for a simple root $\alpha \in \Delta^-$, then $\sigma(\underline{p})$ will be conjugate (under G) to $\underline{b} \oplus \underline{g}_{\sigma_o(\alpha)}$. This implies that σ maps lines of type α in $\mathcal{B}_x$ to lines of type $\sigma_o(\alpha)$ in $\mathcal{B}_x$. Thus we obtain:

<u>Lemma 3</u>: <u>Let</u> x <u>be a subregular nilpotent element in a Lie algebra of type</u> $\Delta = A_r$, D_r <u>or</u> E_6. <u>Then the action of</u> $ZA(x)$ <u>on the dual diagram</u> Δ <u>of the Dynkin curve</u> $\mathcal{B}_x$ <u>is given by the natural homomorphism</u> $ZA(x) \to \mathrm{Aut}(\Delta)$.

8. Deformations of Simple Singularities

8.1. Some Aids.

Lemma 1: *Let $\phi : V \to U$ be a G_m-morphism of type $(d_1,\dots,d_n;w_1,\dots,w_n)$ of two n-dimensional G_m-vector spaces with $\phi(0) = 0$, whose differential $D_0\phi$ at 0 has rank n. Then after a suitable reordering of the weights we will have $d_i = w_i$ for $i = 1,\dots,n$. If the weights $d_i = w_i$ are strictly positive, then ϕ is a polynomial isomorphism.*

Proof: Along with ϕ, its differential $D_0\phi$ is also equivariant with respect to G_m. The equality of the weights and their multiplicities follows then from the fact that $D_0\phi$ induces a G_m-isomorphism of V and U.

Now let d_i, $i = 1,\dots,n$ be strictly positive and let $0 < \lambda_1 < \dots < \lambda_s$ be the distinct values of the weights of G_m on V and U ordered by size. Let V_j resp. U_j denote the G_m-eigenspace in V resp. U of weight λ_j; we then have $\dim V_j = \dim U_j = \#\{i \in \{1,\dots,n\} \mid d_i = \lambda_j\}$. If $v \in V$, and v_l, $l = 1,\dots,s$ is the projection of v to V_l, then because of the ordering of the λ_j, the projection u_j of $\phi(v)$ to U_j has the form

$$u_j = m_j(v_j) + q_j(v_1,\dots,v_{j-1}) ,$$

where $m_j : V_j \to U_j$ is linear and $q_j : V_1 \oplus \dots \oplus V_{j-1} \to U_j$ is a G_m-equivariant morphism of rank 0 at 0. The differential of ϕ is the direct sum

$$D_0\phi = \bigoplus_{j=1}^{s} m_j .$$

Therefore all the m_j must be invertible, and the system

$$v_j = m_j^{-1}(u_j - q_j(v_1,\dots,v_{j-1})) , \; j = 1,\dots,s$$

gives inductively a polynomial inverse of ϕ .

Lemma 2: Let H be a linearly reductive group, which operates regularly and linearly on the vector spaces $V \cong \mathbb{A}^m$ and $U \cong \mathbb{A}^n$. Let $\phi : V \to U$ be an H-equivariant morphism with $\phi(0) = 0$, whose differential at 0 has rank s . Moreover, let there be an isomorphism from G_m to a subgroup of H , through which G_m operates on V with strictly positive weights. Then there is a decomposition of V and U into direct H-summands $V = V_o \oplus B$ and $U = U_o \oplus B$ with $\dim B = s$, as well as an H-equivariant automorphism α of V such that $\phi \circ \alpha : V_o \oplus B \to U_o \oplus B$ will take the form

$$(v,b) \longmapsto (\psi(v,b),b),\ v \in V_o\ ,\ b \in B\ .$$

Proof: The H-equivariance of ϕ implies that of $D_o\phi$. Therefore the image B of $D_o\phi$ in U resp. the kernel V_o of $D_o\phi$ in V are H-stable subspaces to which complementary H-moduls U_o resp. B' exist, because H is linearly reductive. The restriction of $D_o\phi$ to B' is a linear H-isomorphism onto B . We therefore identify B and B' . Let π_o and π_1 denote the canonical projection of V to V_o and B , and without fear of confusion, also the projections of U to U_o and B . Let $\beta : V \to V = V_o \oplus B$ be defined by $\beta(v) = \pi_o(v) \oplus \pi_1 \circ \phi(v)$. Then β has maximal rank m at 0 and is by Lemma 1 an H-isomorphism. Obviously $\phi \circ \beta^{-1}$ has the desired form.

Lemma 3: Let $\phi : V \to U$ be a G_m-morphism of type $(d_1,\dots,d_n;d_1,\dots,d_n)$ of two n-dimensional G_m-vector spaces, whose fiber $\phi^{-1}(0)$ is zero-dimensional. If the weights $d_1,\dots,d_n$ are all strictly positive, then ϕ is an isomorphism.

Proof: If ϕ was not an isomorphism then the rank s of $D_o\phi$ would be less than n (Lemma 1, it follows from our situation that $\phi(0) = 0$). By Lemma 2 we can restrict our attention to the case $s = 0$. So let $s = 0$ and without loss of generality let d_1 be the smallest of the weights d_i , $i = 1,\dots,n$. Since rank $D_o\phi$

is zero and $d_1 \leq d_i$, $i = 2,\dots,n$, any coordinate component ϕ_1 of ϕ with degree d_1 must be identically zero. However, then $\phi^{-1}(0)$ is positive dimensional, in contradiction to our assumption.

8.2. The Subregular Deformations.

Let G be a simple adjoint group of rank r with Lie algebra $\underline{g}$ and adjoint quotient $\gamma : \underline{g} \to \underline{h}/W$ (cf. 3.12). Moreover let $\mathrm{char}(k) = 0$ or $> 4\,\mathrm{Cox}(G) - 2$ and let (x,h,y) be an sl_2-triplet for a subregular nilpotent element $x \in \underline{g}$ (7.2). Then $S = x + \underline{z}_{\underline{g}}(y)$ is a transverse slice to the G-orbit of x on which the product $G_m \times C$ of G_m and the reductive centralizer C of x acts (cf. 7.5). Under the assumptions on $\mathrm{char}(k)$, the group $C \times G_m$ is linearly reductive (1.5,7.5). We will identify the $C \times G_m$-module $\underline{z}_{\underline{g}}(y)$ with S using the translation $z \mapsto x + z$. The restriction $\delta : S \to \underline{h}/W$ of the morphism γ to S is flat (7.4 Corollary) and gives rise to a $C \times G_m$-deformation (cf. 2.5, 7.4, 7.5) of the special fiber $X_o = \delta^{-1}(\bar{0})$ stabilized by $C \times G_m$. We know (7.4, 6.4) that X_o has a single isolated singularity at x which is a rational double point of the type associated with G . In the next sections we will study $\delta : S \to \underline{h}/W$ as a deformation of X_o .

Since the simple singularities have embedding dimension 3 (6.1), and since $\dim S = r + 2$, and $\dim \underline{h}/W = r$, the rank of γ at x is equal to $r - 1$ (for another derivation, see 8.3). Therefore S and $\underline{h}/W \cong \mathbb{A}^r$ decompose into direct $C \times G_m$-summands $S = M \oplus B$ resp. $\underline{h}/W = E \oplus B$ with $\dim B = r - 1$, $\dim E = 1$, $\dim M = 3$, so that by 8.1 Lemma 2 , we may suppose $\delta : S \to \underline{h}/W$ to be of the form $\delta(m,b) = (\delta_E(m,b),b)$, $m \in M$, $b \in B$. The quasihomogeneous type of δ is $(d_1,\dots,d_r;w_1,\dots,w_{r+2})$ with the values given in 7.4 for which $d_i = w_i$ for $i = 1,\dots,r-1$, and $w_j < d_r$ for $j = 1,\dots,r+2$. Therefore the component δ_r of δ of type $(d_r;w_1,\dots,w_{r+2})$ is the unique singular component at x and is equal to δ_E . So G_m operates on B with the weights $d_1 = w_1,\dots,d_{r-1} = w_{r-1}$ and on M with weights w_r,w_{r+1},w_{r+2} . Since the C-action is trivial on $\underline{h}/W$, the group C also operates trivially on E and B .

As a direct result, X_o will be defined as a hypersurface in M by a C-invariant polynomial $f = \delta_r|_M$ of quasihomogeneous type $(d_r;w_r,w_{r+1},w_{r+2})$. As is shown by a look at the list of 7.4 Proposition 2, the values $(d_r;w_r,w_{r+1},w_{r+2})$ are the same for an inhomogeneous Dynkin diagram Δ and its associated homogeneous diagram ${}_h\Delta$. This corresponds to the fact that the singularity of X_o at x is in this case of type ${}_h\Delta$ (cf. 6.4).

By 6.4 the singularity of X_o at x is equivalent (after Henselization) to a singularity defined by one of the <u>normal forms</u>

$X^{r+1} + YZ$	A_r
$X^{r-1} + XY^2 + Z^2$	D_r
$X^4 + Y^3 + Z^2$	E_6
$X^3Y + Y^3 + Z^2$	E_7
$X^5 + Y^3 + Z^2$	E_8

As can be verified immediately, these normal forms are also quasihomogeneous of the type $(d_r;w_r,w_{r+1},w_{r+2})$ as were the polynomials $f = \delta_r|_M$ for the corresponding Lie algebra $\underline{g}$. In fact it will be shown that f can be transformed into the corresponding normal form by composition with an appropriate G_m-isomorphism (8.3 Lemma 3). However, this result is not required in the later sections.

8.3. An alternative Identification of the Subregular Singularities.

Let G , $\underline{g}$, $\gamma : \underline{g} \to \underline{h}/W$, x , S , δ and char(k) be as in 8.2. We wish to derive here the fact that $\mathrm{rank}_x\delta = r-1$ in a way different from 8.2 and without using the results of 6.4. This will lead to an alternative to 6.4 for identifying the singularities of $X_o = \delta^{-1}(\bar{0})$ at x . That this is possible was asserted by Brieskorn in [Br4] in outlining the proof of his main result (cf. 8.7). The following sections will be logically independent of this one. On the other hand this section may replace sections 6.3 and 6.4 assuming the somewhat stronger restrictions on char(k) .

Lemma 1: Let $\phi : \mathbb{A}^{r+2} \to \mathbb{A}^r$ be a quasihomogeneous flat morphism of type $(d_1,\ldots,d_r;w_1,\ldots,w_{r+2})$ with the following properties:

i) The weights w_j, $j = 1,\ldots,r+2$ and the degrees d_i, $i = 1,\ldots,r$, are strictly positive, $w_i = d_i$ for $i = 1,\ldots,r-1$, and $w_j < d_r$ for $j = 1,\ldots,r+2$.

ii) If $f \in k[X,Y,Z]$ is a quasihomogeneous polynomial of type $(d;w_r,w_{r+1},w_{r+2})$ whose associated variety $\operatorname{Spec} k[X,Y,Z]/(f)$ is not smooth and has an isolated singularity at 0, then $d \geq d_r$.

iii) The (schematic) fiber $\phi^{-1}(0)$ has an isolated singularity at 0. Then ϕ has rank $r-1$ at 0.

Proof: Because of the inequality $w_j < d_r$ for all weights w_j, the component ϕ_r of ϕ with degree d_r is necessarily singular at 0, and so rank $D_0\phi \leq r-1$. We now assume that ϕ has rank $s < r-1$ at 0. By 8.1 Lemma 2 we can reduce the situation to the case when $s = 0$. Without loss of generality, let d_1 be the smallest degree among the d_i, $i = 1,\ldots,r$. Then a coordinate component ϕ_1 of ϕ with degree d_1 can not depend on any coordinate of $\mathbb{A}^{r+2}$ whose weight is equal to $w_i = d_i$ for $i = 1,\ldots,r-1$, since these weights are greater or equal to d_1, and since ϕ must have rank 0 at 0. By assumption (ii) and the fact that $d_1 < d_r$, it follows that the hypersurface $\phi_1^{-1}(0)$ in $\mathbb{A}^{r+2} = \mathbb{A}^{r-1} \times \mathbb{A}^3$ has a singular set of dimension $\geq r$. Now $\phi^{-1}(0)$ is obtained from $\phi_1^{-1}(0)$ by intersecting it with $(r-1)$ other hypersurfaces $\phi_2 = \ldots = \phi_r = 0$, given by the other coordinate components ϕ_i of ϕ. Since ϕ is flat the ϕ_i, $i = 1,\ldots,r$ form a regular sequence. The singular set of $\phi^{-1}(0)$ therefore has at least dimension 1, in contradiction to assumption (iii).

Lemma 2: Let $f \in k[X,Y,Z]$ be a quasihomogeneous polynomial of type $(d;w_r,w_{r+1},w_{r+2})$, whose variety $\operatorname{Spec} k[X,Y,Z]/(f)$ has an isolated singularity (and is not smooth). If the values $(d_r;w_r,w_{r+1},w_{r+2})$ correspond to the values of a

Dynkin diagram in 7.4 Proposition 2, then either $d \geq d_r$, or $\Delta = C_4$, D_5 and $d \geq 12 = d_r - 4$.

Proof: For the proof we need consider only homogeneous diagrams Δ (values for Δ = values for ${}_h\Delta$). By the isolatedness assumption, every coordinate must appear in the polynomial f, in particular the coordinate Z of highest weight w_{r+2}. Therefore it suffices to discuss those degrees d with $d_r > d > w_{r+2}$. For the individual types Δ we list the quasihomogeneous type $(d_r; w_r, w_{r+1}, w_{r+2})$ from 7.4 Proposition 2, and the polynomials of type $(d; w_r, w_{r+1}, w_{r+2})$ with $d_r > d > w_{r+2}$ containing Z; as one easily sees one can normalize f without loss of generality such that the coefficients of the monomials are 1. The nonisolatedness of the singularities of Spec $k[X,Y,Z]/(f)$ will be obvious in all cases considered.

E_6 $(24;6,8,12)$: $ZX + X^3$, ZY

E_7 $(36;8,12,18)$: ZX, ZY, ZX^2

E_8 $(60;12,20,30)$: ZX, ZY, ZX^2

A_r $(2r+2;2,r+1,r+1)$: ZX^m, $m = 1,\dots,[\frac{r}{2}]$

D_{2n} $(8n-4;4,4n-4,4n-2)$: ZX^m, $Z(Y+X^{n-1})$

D_{2n+1} $(8n;4,4n-2,4n)$: $X^m(Z+X^n)$ with $m = 1,\dots,n-2$; and
$X^{n-1}(Z+X^n) + Y^2$ with $n > 2$;

For D_5 (and so C_4), $X(Z+X^2)+Y^2$ has an isolated singularity of type A_1, and the quasihomogeneous type of f is $(12;4,6,8)$.

Proposition 1: The morphism $\delta : S \to \underline{h}/W$ has rank $r-1$ at x.

Proof: We use Lemma 1. Property (ii) follows from Lemma 2 for $\Delta \neq C_4$, D_5. Properties (i) and (iii) are satisfied because of 7.4 Proposition 2 and 5.5. The cases C_4 and D_5, in which δ has the type $(4,8,(10),12,16;4,8,(10),12,4,6,8)$, can be reduced by arguments similar to those at the end of the proof of Lemma 1, using 8.1 Lemma 2, to the situation of a morphism $\phi : \mathbb{A}^4 \to \mathbb{A}^2$ of type $(12,16;4,6,8,12)$ which only has an isolated singularity in $\phi^{-1}(0)$ when $\mathrm{rank}_0\phi = 1$ (when $\mathrm{rank}_0\phi$ is zero the coordinate axis of weight 12 lies in the singular set of $\phi^{-1}(0)$).

Remark: By no means can one conclude directly from the isolatedness of the singularity of $\delta^{-1}(0)$, the flatness of δ, and the equality of the values of $r-1$ degrees and weights that $\mathrm{rank}_x\delta = r-1$. Consider for example the flat morphism $\phi : \mathbb{A}^4 \to \mathbb{A}^2$, $(x,y,z,u) \mapsto (x^2+y^2+z^2, y^4+z^4+u^2)$ of type $(2,4;1,1,1,2)$, with $\mathrm{rank}_0\phi = 0$. It can be easily verified that $\phi^{-1}(0)$ has an isolated singularity at 0.

From the statement that $\mathrm{rank}_x\delta = r-1$ it follows as in 8.2 that $X_0 = \delta^{-1}(\bar{0})$ is defined by a quasihomogeneous polynomial $f : \mathbb{A}^3 \to \mathbb{A}^1$ of type $(d_r;w_r,w_{r+1},w_{r+2})$. An identification of the singularity of X_0 at x independent of 6.4 is given in the following proposition (in which it suffices to assume $\mathrm{char}(k) \neq 2$).

Proposition 2: Let $f \in k[X,Y,Z]$ be a quasihomogeneous polynomial of type $(d_r;w_r,w_{r+1},w_{r+2})$ with values corresponding to a Dynkin diagram Δ. If $\mathrm{Spec}\, k[X,Y,Z]/(f)$ has an isolated singularity, then there is a G_m-equivariant automorphism α of $\mathbb{A}^3$ such that $f \circ \alpha$ has the normal form of the rational double point corresponding to Δ (resp. ${}_h\Delta$).

Proof: The claims of quasihomogeneity and isolatedness lead to conditions on the monomials appearing in f and their coefficients which will allow us to define certain transformations into normal forms. In so far as is possible we will normalize the coefficients of the monomials to 1.

1) A_{2n} : The type is $(4n+2;2,2n+1,2n+1)$, so only the monomials X^{2n+1}, Y^2, Z^2, YZ can appear in f. By the linear equivalence of all nondegenerate quadrics in Y and Z, f is equivalent to $X^{2n+1} + YZ$.

2) A_{2n-1} : Type $(4n;2,2n,2n)$, possible monomials are X^{2n} , X^nY , X^nZ , Y^2 , YZ , Z^2 . For $n = 1$, f must be a nondegenerate quadratic form which can be linearly tranformed into $X^2 + YZ$. For $n > 2$, the quadratic form appearing in f can be brought linearly to the form $Y^2 + Z^2$. The terms X^nY and X^nZ can be removed by a quasihomogeneous transformation $Y \mapsto Y + cX^n$ or $Z \mapsto Z + dX^n$ for appropriate $c,d \in k$ ("completing the square"). Now transform $Y^2 + Z^2$ into YZ .

3) D_{2n+1} , $n \geq 2$: The type $(8n;4,4n-2,4n)$ allows the monomials X^{2n} , XY^2 , Z^2 , ZX^n . By isolatedness Z^2 must appear. A monomial ZX^n can be removed by completing the square $Z \mapsto Z + cX^n$, for a suitable $c \in k$. The monomials X^{2n} and XY^2 now appear because of the isolatedness condition.

4) D_{2n} , $n > 2$: Possible monomials for the type $(8n-4;4,4n-4,4n-2)$ are X^{2n-1} , XY^2 , YX^n , Z^2 . The monomial YX^n can be removed by completing the square $Y \mapsto Y + cX^{n-1}$, $c \in k$. For the rest see (3) .

5) D_4 : Type $(12;4,4,6)$. By isolatedness f has the form $Z^2 + q(X,Y)$ where $q(X,Y)$ is a cubic form which decomposes into three different linear factors. Such cubic forms are all linearly equivalent to $X^3 + XY^2$.

6) E_6 : The type $(24;6,8,12)$ allows monomials X^4 , Y^3 , Z^2 , ZX^2 . By isolatedness Z^2 appears. ZX^2 can be transformed away by $Z \mapsto Z + cX^2$ for appropriate $c \in k$, and the isolatedness then implies that $f \sim X^4 + Y^3 + Z^2$.

7) E_7 resp. E_8 : Type $(36;8,12,18)$ resp. $(60;12,20,30)$. In these cases only three monomials can appear and must appear by isolatedness.

8.4. The Simplicity of the Subregular Singularities.

Let G be a simple adjoint group of type Δ with Lie algebra $\underline{g}$, $\mathrm{char}(k) = 0$ or $> 4\,\mathrm{Cox}(G) - Z$, (x,h,y) an sl_2-triplet for a subregular nilpotent element $x \in \underline{g}$, and $\delta : S \to \underline{h}/W$ the restriction of the adjoint quotient of $\underline{g}$ to the transverse slice $S = x + \underline{z}_{\underline{g}}(y)$. We know from 6.4 or 8.3 that $X_o = \delta^{-1}(\bar{0})$ is a rational double point of type Δ (resp. ${}_h\Delta$) when Δ is homogeneous (resp. inhomogeneous). If Δ

is homogeneous, then by definition (cf. 6.1) X_o is simple of the <u>same</u> type. In this section we will establish the corresponding result for inhomogeneous Δ. For this we have to define a group Γ of symmetries on X_o (cf. 6.2).

Assume from now on that Δ is inhomogeneous. Since the reductive centralizer $C(x)$ of x operates on X_o, the following makes sense: Define a group $\Gamma \subset \mathrm{Aut}(X_o)$ by

$$\Gamma = \begin{cases} C(x) & \text{if } \Delta = C_r, F_4, G_2 \\ \{1,s\} & \text{if } \Delta = B_r. \text{ Here } s \in C(x)\setminus C(x)^o \text{ is} \\ & \text{a nontrivial involution.} \end{cases}$$

In the case $\Delta = B_r$ the group Γ is well defined up to conjugation by $C(x)^o = G_m$.

<u>Theorem</u>: <u>The couple</u> (X_o,Γ) <u>is a simple singularity of type</u> Δ.

<u>Proof</u>: We use Proposition 6.2. From 6.4 or 8.3 we know that X_o is a rational double point of type ${}_h\Delta$. Proposition 7.5 tells us that Γ is isomorphic to $AS(\Delta)$ and that the action of Γ on the dual diagram ${}_h\Delta$ of the minimal resolution coincides with the associated action of $AS(\Delta)$. Thus it only remains to prove that the action of Γ on X_o is free (in the sense of 6.2). More generally let c be an element of $C(x)$, and suppose $c\cdot v = v$ for an element $v \in X_o$, $v \neq x$. Then c lies in the centralizer $Z_G(v)$ of the regular nilpotent element v. According to 7.5 the group $Z_G(v)$ contains only unipotent elements, whereas $C(x)$ consists of semisimple elements. We thus get $c = 1$, which implies the freeness of Γ on X_o. Therefore Proposition 6.2 applies and proves the theorem.

As in 8.3, a purely algebraic proof of the theorem is possible, making no use of the resolution of the singularities (6.4). For this we assume $\delta : S \to \underline{h}/W$ is of the form $\delta : M \oplus B \to E \oplus B$, $\delta(m,b) = (\delta_r(m,b),b)$, as described in 8.2. The singularity $X_o = \delta^{-1}(\overline{0})$ is then defined as a hypersurface $\{m \in M \mid f(m) = 0\}$ in M, where

$f = \delta_r|_M$ is a $C(x)$-invariant quasihomogeneous polynomial of type $(d_r; w_r, w_{r+1}, w_{r+2})$. Without any risk of confusion we denote $C(x)$ by C and consider M as a $C \times G_m$-submodule of $\underline{z}_{\underline{g}}(y)$.

Lemma: The G_m-eigenspaces $M(i)$ of weight $i \in \mathbb{Z}$ are C-submodules of M. The group C operates faithfully on M by transformations of determinant 1.

Proof: The first statement follows trivially from the fact that C and G_m commute. To prove the second statement we consider $\underline{g}$ as an sl_2-module with respect to the triplet (x,h,y). Denote by $\overline{M}$ (resp. $\overline{B}$) the sl_2-submodules generated by M (resp. B). Since $\underline{z}_{\underline{g}}(y) = M \oplus B$ consists of the lowest weight vectors of the irreducible sl_2-submodules in $\underline{g}$, which generate $\underline{g}$ as an sl_2-module, we obtain $\underline{g} = \overline{M} \oplus \overline{B}$. As a subgroup of the adjoint group G the reductive centralizer C has to act faithfully on $\underline{g}$ by transformations of determinant 1. The action of C is trivial on B (cf. 8.2) as well as on $\overline{B}$, because C commutes with sl_2. Therefore C has to act faithfully on $\overline{M}$ by transformations of determinant 1. To see that C acts faithfully on M it suffices to remark again that C and sl_2 commute. For the statement on the determinant a finer analysis is required. Denote by $\overline{M}_n$ that sl_2-isotypical component in $\overline{M}$ whose irreducible summands are n-dimensional. Then $M \cap \overline{M}_n$ coincides with the h-eigenspace of weight $-(n-1)$, i.e. with the G_m-eigenspace $M(n+1)$ of weight $n+1 = 2+(n-1)$ (cf. 7.4 Prop. 1, proof). With respect to C, the module $\overline{M}$ decomposes into a sum of n summands isomorphic to $M \cap \overline{M}_n = M(n+1)$. For an element $c \in C$ we thus have

$$1 = \det c = \det(c|_{\overline{M}}) = \prod_{n \in \mathbb{N}} (\det(c|_{M(n+1)}))^n .$$

We know that this product actually runs over the following three values for n :

$$w_r - 1 \ , \ w_{r+1} - 1 \ , \ w_{r+2} - 1 .$$

A glance at the table of 7.4 Proposition 2 shows these numbers to be odd (for B_r,

C_r , F_4 , G_2). Since the factor commutator group of C is $\mathbb{Z}/2\mathbb{Z}$ ($C = G_m \rtimes \mathbb{Z}/2\mathbb{Z}$, $\mathbb{Z}/2\mathbb{Z}$, $\mathbb{Z}/2\mathbb{Z}$, $\mathfrak{S}_3$ for $\Delta = B_r$, C_r , F_4 , G_2) the formula above reduces to

$$1 = \prod_{n \in \mathbb{Z}} (\det(c_{|M(n+1)}))^n = \prod_{n \in \mathbb{Z}} \det(c_{|M(n+1)}) = \det(c_{|M}) \ .$$

Combined with this lemma, the following proposition yields another proof of the theorem above.

Proposition: Let Δ be a diagram of type B_r , C_r , F_4 , or G_2 , and let C be a group isomorphic to $G_m \rtimes \mathbb{Z}/2\mathbb{Z}$, $\mathbb{Z}/2\mathbb{Z}$, $\mathbb{Z}/2\mathbb{Z}$, or $\mathfrak{S}_3$ respectively. Assume that C acts faithfully on a threedimensional vector space M by transformations of determinant 1 leaving invariant a quasihomogeneous polynomial f of type $(d_r; w_r, w_{r+1}, w_{r+2})$ with weights and degrees corresponding to Δ (7.4 Proposition 2). Assume moreover that C commutes with the induced G_m-action on M and that $X_o = f^{-1}(0)$ has an isolated singularity. For $\Delta \neq B_r$ let $\Gamma = C$, and for $\Delta = B_r$ let $\Gamma = \{1,s\}$, where $s \in G_m \rtimes \mathbb{Z}/2\mathbb{Z}$, $s \notin G_m$, is a nontrivial involution in C . Let $\mathrm{char}(k)$ be good for Δ . Then (X_o, Γ) is a simple singularity of type Δ .

Proof: We will describe the action of C on M , case by case. A comparison with the explicit formulae for the simple singularities (6.2) then gives the result. To obtain the structure of X_o we may apply 8.3 proposition 2; yet we quickly repeat some of the arguments.

1) B_r , $r \geq 2$, ${}_h\Delta = A_{2r-1}$. The polynomial f is quasihomogeneous of type $(4r; 2, 2r, 2r)$. We choose G_m-equivariant coordinates X,Y,Z on M (of weight 2, 2r, 2r). The group C is isomorphic to $G_m \rtimes \mathbb{Z}/2\mathbb{Z}$. Let $\sigma : C \to \mathbb{Z}/2\mathbb{Z}$ be the projection to $\mathbb{Z}/2\mathbb{Z}$ and let s be any involution such that $\sigma(s) = -1$. There is exactly one nontrivial, one-dimensional representation (given by the character σ) and exactly one faithful two-dimensional representation as the orthogonal group O_2 (the two G_m-eigenspaces of such a two-dimensional representation will be permuted

by s , and thus have opposite weights n and $-n$. The faithfulness then implies $n = 1$).

Since C operates faithfully on M , C must operate on the eigenspace $M(2r)$ as the orthogonal group O_2 , and because of the determinant condition $(\det(c_{|M}) = 1$, $c \in C$) must operate on $M(2)$ by σ .

As a result of the quasihomogeneity, at most the following monomials could appear in f : X^{2r} , X^rY , X^rZ , Y^2 , YZ , Z^2 . The quadratic term appearing in f will be non-degenerate and, after a linear coordinate change in $M(2r)$, we may assume this term to be YZ . We then have $C = O_2(YZ)$. A term $X^r(\alpha Y+\beta Z)$, $\alpha,\beta \in k$, is only invariant under C when $\alpha = \beta = 0$. Therefore, after scaling the coordinate X , f will have the form $X^{2r} + YZ$. And after conjugation by an appropiate element of $C^o = SO_2$, the involution s will operate by $X,Y,Z \mapsto -X,Z,Y$.

2) C_r , $r \geq 3$, ${}_h\Delta = D_{r+1}$. The quasihomogeneous type of f is $(4r;4,2r-2,2r)$. Since the group C is isomorphic to $\mathbb{Z}/2\mathbb{Z}$; and since its action is faithful on M and the determinant condition holds, C must operate on two of the three G_m-equivariant coordinates X,Y,Z by -1 .

Now let $r > 3$. Besides the monomials X^r , XY^2 , Z^2 , the monomials ZX^n , for $r = 2n$, and YX^{n+1} , for $r = 2n+1$, can appear in f . By the isolatedness of the singularity of X_o , XY^2 must appear. Therefore C can only operate by $X,Y,Z \mapsto X,-Y,-Z$, and after adjusting the coordinates, f will have the form $X^r + XY^2 + Z^2$.

Now let $r = 3$. Then f has the form $Q(X,Y) + Z^2$, where Q is a cubic form on $M(4)$ which will decompose into three different linear factors. Therefore C cannot act on $M(4)$ by -1 (the form Q would not be invariant). Without loss of generality, C operates on M by $X,Y,Z \mapsto X,-Y,-Z$. Then f has the form $X^3 + XY^2 + Z^2$.

3) G_2 , ${}_h\Delta = D_4$, Type $(12;4,4,6)$, $C \cong \mathfrak{S}_3$. We regard C as the subgroup $\mu_3 \rtimes \mathbb{Z}/2\mathbb{Z} \subset G_m \rtimes \mathbb{Z}/2\mathbb{Z}$, $\mu_3 = \{t \in G_m | t^3 = 1\}$, of the orthogonal group O_2 , whose

natural representation induces the unique irreducible two dimensional representation of C . The restriction of σ (cf. case (1) B_r) to C gives the unique non-trivial homomorphism $C \to G_m$.

By faithfulness and the determinant condition, C operates on $M(4)$ as a subgroup of O_2 , where we may assume $O_2 = O_2(XY)$, and on $M(6)$ by σ . The alternating group $\mathfrak{A}_3 = \mu_3 \subset C$ operates by $X,Y,Z \mapsto tX,t^2Y,Z$ for all t (or t^{-1}) $\in \mu_3$ on M . The polynomial f now has the form $X^3 + Y^3 + Z^2$, since $X^3 + Y^3$ is the unique nondegenerate C-invariant cubic form in X and Y (up to scalar adjustments of the coordinates). One of the involutions in C permutes X and Y .

4) F_4 , ${}_h\Delta = E_6$. Here f has the type $(24;6,8,12)$ and must be of the form $\varepsilon X^4 + Y^3 + \lambda ZX^2 + Z^2$, where $\varepsilon = 0,1$ and $\lambda \in k$ (moreover $\lambda \neq 0$ when $\varepsilon = 0$, and $\lambda \neq \pm 2$ when $\varepsilon = 1$). The G_m-module M decomposes into the eigenspaces $M(6)$, $M(8)$, $M(12)$ corresponding to the coordinates X,Y,Z . The group $C = \mathbb{Z}/2\mathbb{Z}$ must operate on two coordinates by -1 (faithfulness, determinant condition). The only possible solution is $X,Y,Z \mapsto -X,Y,-Z$, which also implies $\lambda = 0$ and $\varepsilon = 1$.

Remark: In the case $\Delta = B_r$, if we would not consider the action of C but only the action of an involution $s \in C$, then this action would not be uniquely determined. The other possibility for s is the operation $X,Y,Z \mapsto X,-Y,-Z$, which is realized by the action of $\mathbb{Z}_{4r}/\mathbb{Z}_{2r}$ on $X_o \cong \mathbb{A}^2/\mathbb{Z}_{2r}$.

8.5. A Supplement to 7.5.

In this section we calculate the reductive centralizers for the subregular nilpotent elements in Lie algebras of type F_4 , E_6 , E_7 , E_8 . The method is close to the algebraic proof of Theorem 8.4. We will use the same notation, and also make use of that part of the proof of Lemma 8.4 which does not depend on the explicit form of C . In the beginning the arguments concerning F_4 and E_6 coincide, since they correspond to the same singularity.

The quasihomogeneous type of f in case $E_6(F_4)$ resp. E_7 resp. E_8 is

(24;6,8,12) resp. (36;8,12,18) resp. (60;12,20,30). Correspondingly the dimensions of the three irreducible sl_2-submodules of $\bar{M} \subset \underline{g}$ are 5,7,11 resp. 7,11,17 resp. 11,19,29. Up to scalar adjustments of the G_m-equivariant coordinates the polynomial f has the form $\varepsilon X^4 + Y^3 + Z^2 + \lambda Z X^2$ (cf. 8.4, case 4) resp. $X^3Y + Y^3 + Z^2$ resp. $X^5 + Y^3 + Z^2$. Thus the group C can only act on M by means of a homomorphism $c \longmapsto (\xi(c),\eta(c),\zeta(c))$ into the group $\mu_4 \times \mu_3 \times \mu_2$ resp. $\mu_9 \times \mu_3 \times \mu_2$ resp. $\mu_5 \times \mu_3 \times \mu_2$. The determinant condition $\det(c_{|\bar{M}}) = \det c = 1$ is equivalent to $1 = \xi^5 \cdot \eta^7 \cdot \zeta^{11} = \xi \cdot \eta \cdot \zeta$ resp. $1 = \xi^7 \cdot \eta^{11} \cdot \zeta^{17} = \xi^{-2} \cdot \eta^2 \cdot \gamma$ resp. $1 = \xi^{11} \cdot \eta^{19} \cdot \zeta^{29} = \xi \cdot \eta \cdot \zeta$.

In the case E_8 this condition can only be satisfied by $\xi = \eta = \zeta = 1$, which implies $C = 1$ (the action of C on M is faithful). For E_7 we obtain $\xi = \eta$ and $\gamma = 1$. In addition the monomial X^3Y is C-invariant. This forces $1 = \eta^4 = \eta$, and $C = 1$. For $E_6(F_4)$ the condition says $\eta = 1$ and $\xi = \eta$. Hence we get two possibilities: either C is isomorphic to $\mathbb{Z}/2\,\mathbb{Z}$, or C is trivial. In the case F_4 we know that C is nontrivial as it permutes transitively the lines of the same type in the Dynkin curve (cf. 7.5). Thus $C = \mathbb{Z}/2\,\mathbb{Z}$. On the other hand, all lines in a Dynkin curve for E_6 have different type, and C conserves the type. Thus $C = 1$ for E_6 .

<u>8.6. The G_m-Structure on the Semiuniversal Deformations of the Simple Singularities.</u>

Let F be a linearly reductive finite subgroup of SL_2 . Then the usual scalar multiplication on $\mathbb{A}^2$ induces a natural action of the multiplicative group G_m on the quotient $\mathbb{A}^2/F$. If $F' \rhd F$ is another subgroup of SL_2 normalizing F , then the action of F'/F on $\mathbb{A}^2/F$ will commute with the G_m-action. Thus all simple singularities aquire a natural G_m-action. According to the descriptions given in 6.1, 6.2, 7.4 and 8.3 this action coincides with the one given by the quasihomogeneous normal form of a rational double point.

<u>Proposition:</u> <u>Let</u> Δ <u>be an irreducible Dynkin diagram and</u> $\xi : X \to (U,o)$ <u>a</u> <u>G_m-semiuniversal G_m-deformation of a simple singularity of type</u> Δ . <u>Assume that</u> char(k) <u>does not divide the Coxeter number of</u> Δ . <u>Then we have:</u>

i) dim $U = r = \text{rank } \Delta$

ii) The weights of G_m on U are $d_1, \dots, d_r$, where $d_1, \dots, d_r$ are the degrees multiplied by two of the invariant polynomials on a simple Lie algebra of type Δ.

Proof: First assume that Δ is homogeneous and that the corresponding simple singularity is given by a quasihomogeneous polynomial f of type $(d_r; w_r, w_{r+1}, w_{r+2})$ with values d_i, w_j as in 7.4 Proposition 2. According to 2.7 we only need to show that the polynomial $\sum_{i=1}^{r} T^{d_i}$ satisfies the equation for $P_U(T)$ given there (in 2.7). This can be done easily by means of the table in 7.4 Propositon 2. Now let Δ be inhomogeneous and (X_o, Γ) a simple singularity of type Δ defined by a Γ-invariant quasihomogeneous polynomial f of type $(d_r; w_r, w_{r+1}, w_{r+2})$, where the weights and degrees are as in 7.4 Proposition 2. To calculate the weights of G_m on U we use a $\Gamma \times G_m$-semiuniversal $\Gamma \times G_m$-deformation $\eta : Y \to (V, o)$ of X_o, and we identify U with the fixed-point space V^Γ (cf. 2.5, 2.6). The $\Gamma \times G_m$-module V is isomorphic to a $\Gamma \times G_m$-complement in $k[X,Y,Z]$ of the Jacobian ideal $I(f)$, which is generated by the partial derivatives $\frac{\partial f}{\partial X}, \frac{\partial f}{\partial Y}, \frac{\partial f}{\partial Z}$ (cf. 2.5, 2.7, under our assumptions on char(k) we have $f \in I(f)$). To obtain the correct weights we have to regard $k[X,Y,Z]$ as twisted by the G_m-character of weight d_r. Thus a monomial $X^i Y^j Z^l$ will have weight $d_r - iw_r - jw_{r+1} - lw_{r+2}$ (cf. the construction of V in 2.5). In the following case-by-case discussion we assume that f and Γ have the form as described in 6.2 (because of 2.8 Corollary this is no restriction; moreover, the proof of 8.4 Proposition shows that the linearity of Γ forces f to have this form).

1) B_r. The polynomial $f = X^{2r} + YZ$ is of type $(4r; 2, 2r, 2r)$ and a complement to $I(f)$ is spanned freely over k by $1, X, X^2, \dots, X^{2r-2}$. The nontrivial element of $\Gamma = \mathbb{Z}/2\mathbb{Z}$ operates by $X^i \mapsto (-X)^i$. Thus $U = V^\Gamma$ is spanned by the monomials $1, X^2, X^4, \dots, X^{2r-2}$, whose G_m-weights are $4r, 4r-4, \dots, 8, 4$.

2) C_r . Here $f = X^r + XY^2 + Z^2$ is of type $(4r;4,2r-2,2r)$. A complement to $I(f)$ will be freely spanned by $1,X,\ldots,X^{r-1},Y$, which are all invariant under Γ except for Y . As a result we obtain the G_m-weights $4r,4r-4,\ldots,8,4$ on U .

3) F_4 . Now $f = X^4 + Y^3 + Z^2$ is of type $(24;6,8,12)$. A complement to $I(f)$ is given by the monomials $1,X,X^2,Y,YX,YX^2$, which, except for X and XY , are invariant under Γ . Hence G_m acts on U by the weights $24,12,16,4$.

4) G_2 . We assume f to be of the form $X^3 + Y^3 + Z^2$, of type $(12;4,4,6)$. A complement to $I(f)$ is given by $1,X,Y,XY$, which decomposes under the action of $\Gamma = \mathfrak{S}_3$ into the sum of the two-dimensional representation and two trivial summands spanned by 1 and XY . Thus the weights of G_m on U are 12 and 4 .

Remark: Let X_o be a rational double point and ϕ an automorphism of X_o . In general, the condition that ϕ permutes exceptional components of the minimal resolution of X_o is not equivalent to the condition that ϕ operates on the basis V of a semiuniversal deformation in a nontrivial way. For example, in the case $X^{2r} + YZ = 0$, the automorphism $X,Y,Z \mapsto -X,Y,Z$ operates nontrivially on V and permutes no exceptional components, whereas the automorphism $X,Y,Z \mapsto X,Z,Y$ acts trivially on V and permutes exceptional components.

8.7. The Semiuniversality of the Subregular Deformations.

We have prepared all the tools we need to prove our main result.

Theorem: Let G be a simple group with Lie algebra $\underline{g}$, Dynkin diagram Δ and adjoint quotient $\gamma : \underline{g} \to \underline{h}/W$. Let $\operatorname{char}(k) = 0$ or $> 4\operatorname{Cox}(G) - 2$, and let (x,h,y) be an sl_2-triplet for a subregular nilpotent element $x \in \underline{g}$. Then the restriction $\delta : S \to (\underline{h}/W , \bar{0})$ of γ to the transverse slice $S = x + \underline{z}_{\underline{g}}(y)$ is a G_m-semiuniversal G_m-deformation of a simple singularity of type Δ .

Proof: Because of our assumptions about $\operatorname{char}(k)$ we may assume that G is adjoint (cf. 3.13). For Δ of type B_r, C_r, F_4 or G_2 we choose a subgroup Γ of the reductive centralizer $C(x) = Z_G(x) \cap Z_G(y)$ as in 8.4. Then δ is invariant with respect to the natural action of Γ on S. If Δ is homogeneous (resp. inhomogeneous) we know from 6.4 or 8.3 (resp. 8.4) that $X_o = \delta^{-1}(\bar{o})$ (resp. (X_o,Γ)) is a simple singularity of type Δ. Let $\xi : X \to (U,o)$ be a G_m-semiuniversal G_m-deformation of X_o resp. (X_o,Γ). Then there is G_m-morphism (ϕ,Φ) from the deformation δ to ξ:

$$\begin{array}{ccccc} & & X_o & & \\ & \swarrow & & \searrow & \\ S & & \xrightarrow{\quad\Phi\quad} & & X \\ \delta\big\downarrow & & & & \big\downarrow\xi \\ \underline{h}/W & & \xrightarrow{\quad\phi\quad} & & U \end{array}$$

Since the G_m-weights on S, $\underline{h}/W$, X, U are strictly positive, the formal morphisms ϕ and Φ are actually polynomial and globally defined. Moreover $\underline{h}/W$ and U are isomorphic G_m-modules (cf. 7.4, 8.6). If we can show that $\phi^{-1}(0) = \bar{0}$, then, by 8.1 Lemma 3, the G_m-morphism ϕ must be an isomorphism. Since the diagram above is cartesian, every fiber of δ over $\phi^{-1}(0)$ is isomorphic to X_o. However by 6.5 such a fiber occurs only over $\bar{0}$. Therefore $\phi^{-1}(0) = \bar{0}$, and ϕ is an isomorphism. This in turn implies that (ϕ,Φ) is an isomorphism, which proves the theorem.

Corollary: Let G be a simple group of type Δ and $\underline{g}$ its Lie algebra. Let S be a transverse slice to the orbit of a subregular nilpotent (resp. unipotent) element $x \in \underline{g}$ (resp. $x \in G$) meeting this orbit only at x. For Δ of type B_r, C_r, F_4, G_2 assume that S is stable under the image of a section of the projection $Z_G(x) \to Z_G(x)/Z_G(x)^o \cong AS(\Delta)$. Let $\operatorname{char}(k) = 0$ if $x \in G$, and $\operatorname{char}(k) = 0$ or $>$ $> 4\,\operatorname{Cox}(G) - 2$ if $x \in \underline{g}$. Then the restriction δ of the adjoint quotient $\gamma : \underline{g} \to \underline{h}/W$ (resp. $\chi : G \to T/W$) to S induces a formal semiuniversal deformation of a simple singularity of type Δ.

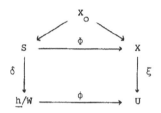

<u>Proof:</u> Use the comparison results of 5.1 Lemma 3 and 2.8 Corollary as well as 7.5 Lemmata 1.4 (Δ inhomogeneous) and 3.15 (if $x \in G$) to reduce to the theorem.

<u>Remarks:</u> 1) For Δ of type A_r, D_r, or E_r the theorem was conjectured in essence by Grothendieck and subsequently proved by Brieskorn in the category of complex analytic space germs. The only published account is the sketch in [Br4]. We have followed this by making use of the quasihomogeneous structure of δ.

2) In the case $\Delta = A_n$ one can prove the theorem directly by calculating the restriction of the elementary symmetric polynomials of sl_{n+1} to a transversal slice (cf. [A1]).

3) Let Δ_o be an inhomogeneous diagram and Δ the associated homogeneous diagram, let G_o, G be the corresponding simple adjoint groups of type Δ_o, Δ and let δ_o and δ be the subregular deformations in the Lie algebras $\underline{g}_o$ and $\underline{g}$. By the semiuniversality of δ, δ_o is induced by δ:

$$\begin{array}{ccc} S_o & \xrightarrow{\ i\ } & S \\ {\scriptstyle \delta_o}\big\downarrow & & \big\downarrow{\scriptstyle \delta} \\ \underline{h}_o/W_o & \xrightarrow{\ j\ } & \underline{h}/W \end{array}$$

It is natural to ask if one can obtain the morphism (i,j) from δ_o to δ by means of a homomorphism $\rho : G_o \to G$ resp. $D_e\rho : \underline{g}_o \to \underline{g}$ which sends a suitable "subregular" transverse slice $S_o \subset \underline{g}_o$ to a similar slice S in $\underline{g}$, and realizes $j : \underline{h}_o/W_o \to \underline{h}/W$ by the natural map of quotients $\underline{g}_o/G_o \to \underline{g}/G$. For simple reasons such a ρ can not exist. If $x \in \underline{g}_o$ is subregular nilpotent and $C(x)$ is the reductive centralizer of x, then $\rho(C(x))$ is contained in the reductive centralizer of $D_e\rho(x)$, which is trivial (or $\cong G_m$ for A_{2r-1}) by 7.5 Lemma 4 when $D_e\rho(x)$ is subregular in $\underline{g}$. However, $C(x)$ is nontrivial (or $\cong G_m \rtimes \mathbb{Z}/2\mathbb{Z}$ for B_r) and ρ is injective (by simplicity and adjointness). (In the cases B_r and C_r, it is also easy to conclude from the dimensions of the fundamental representations (in char(k) = 0) that no (nontrivial) homomorphism $\underline{g}_o \to \underline{g}$ can exist. For F_4 and G_2 there are such homomorphisms, which are unique up to conjugation (cf. [Dy] Th. 11.1), but which do not send subregular elements to subregular elements. The nilpotent class 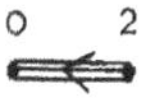

in G_2 goes to the class 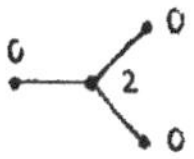

in D_4 and 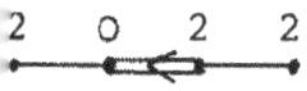

goes to 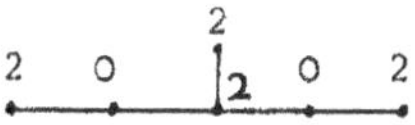

.)

In the next section we will see how the "missing" reductive centralizer in G can be found in the full automorphism group $\mathrm{Aut}(\underline{g})$.

8.8. Outer Automorphisms and Associated Symmetries.

Let Δ be a Dynkin diagram of type A_{2r-1} , $r > 1$, D_r or E_6 , and let G be a simple adjoint group of type Δ with Lie algebra $\underline{g}$. We assume char(k) = 0 or $> 4\,\mathrm{Cox}(G) - 2$. Let (x,h,y) be an sl_2-triplet for a subregular nilpotent element $x \in \underline{g}$, and let $\delta : S \to \underline{h}/W$ be the restriction of the invariant morphism $\gamma : \underline{g} \to \underline{h}/W$ to the transverse slice $S = x + \underline{z}_{\underline{g}}(y)$. The space S is naturally equipped with an action of $CA(x) \times G_m$, where $CA(x)$ is the outer reductive centralizer of x (cf. 7.6). The group $\mathrm{Aut}(\Delta) \cong \mathrm{Aut}(\underline{g})/G$ operates naturally on the quotient $\underline{g}/G \cong \underline{h}/W$, and $\mathrm{Aut}(\Delta)$ may also be considered as a subgroup of $CA(x)$ (cf. 7.6). With respect to the resulting actions of $\mathrm{Aut}(\Delta)$ on S and $\underline{h}/W$, the morphism $\delta : S \to \underline{h}/W$ is equivariant. As a result, there is an action of $\mathrm{Aut}(\Delta)$ on the fiber $X_o = \delta^{-1}(\bar{0})$. Our aim is to show that $(X_o , \mathrm{Aut}(\Delta))$ is a simple singularity of type Δ_o , where Δ_o is the unique inhomogeneous Dynkin diagram with $h\Delta_o = \Delta$ and $AS(\Delta_o) = \mathrm{Aut}(\Delta)$.

We need some preparation. Let T be a maximal torus of G and $B \supset T$ a Borel subgroup containing T . We denote the corresponding Lie algebras by $\underline{h}$ and $\underline{b}$. From LIE VIII § 5 n° 1,2,3 and [St O] we get the following commutative diagram

$$\begin{array}{ccccccccc}
1 & \longrightarrow & G & \longrightarrow & \mathrm{Aut}(\underline{g}) & \longrightarrow & \mathrm{Aut}(\Delta) & \longrightarrow & 1 \\
 & & \uparrow & & \uparrow & & \| & & \\
1 & \longrightarrow & N(T) & \longrightarrow & \mathrm{Aut}(\underline{g};\underline{h}) & \longrightarrow & \mathrm{Aut}(\Delta) & \longrightarrow & 1 ,
\end{array}$$

in which $N(T)$ is the normalizer of T in G and $\mathrm{Aut}(\underline{g};\underline{h})$ denotes the automorphisms of $\underline{g}$ which stabilize $\underline{h}$. The quotient of $\mathrm{Aut}(\underline{g};\underline{h})$ by $T \subset N(T)$ is isomorphic to the automorphism group $\mathrm{Aut}(\Sigma)$ of the root system Σ of T in G. Moreover, the rows of the diagram are exact. The commutativity of the diagram above implies the equivariance of the identification $\underline{h}/W \cong \underline{g}/G$ (induced by the inclusion $\underline{h} \subset \underline{g}$, cf. 3.12) with respect to the naturally induced actions of $\mathrm{Aut}(\Delta)$.

Lemma 1: The group $\mathrm{Aut}(\Delta)$ operates linearly on $\underline{h}/W$ with respect to a suitable choice of homogeneous generators of $k[\underline{h}]^W$. This action commutes with the natural G_m-action on $\underline{h}/W$, and $\mathrm{Aut}(\Delta)$ acts trivially on the G_m-eigenspace of heighest weight.

Proof: The linearity of the $\mathrm{Aut}(\Delta)$-action and the commutativity with G_m are obtained in the following way. Because of the linearity of the $\mathrm{Aut}(\underline{g};\underline{h})$-action, the group $\mathrm{Aut}(\Delta)$ respects the natural grading $A = k \oplus A(1) \oplus A(2) \oplus \dots$ of $A = k[\underline{h}]^W$. This already implies the commutativity with G_m. Every homogeneous component $A(i)$ contains an $\mathrm{Aut}(\Delta)$-stable submodule $A'(i)$, which consists of those elements which can be generated algebraically from the elements in $\bigoplus_{j<i} A(j)$. A linear basis of an $\mathrm{Aut}(\Delta)$-module complement to $A'(i)$ in $A(i)$ will generate the sought-for homogeneous invariants of degree i.

For the cases A_{2r-1}, D_{2n+1}, E_6, the element $-\mathrm{id}_{\underline{h}}$ doesn't lie in the Weylgroup, but rather in $\mathrm{Aut}(\Sigma)$ (cf. LIE IV, V, VI Planches). Therefore $\mathrm{Aut}(\Delta)$ operates trivially on the homogeneous invariants of even degree, in particular on that of the highest degree (cf. 7.4 Proposition 2, note that the values d_i, $i=1,\dots,r$, given there are twice the degrees of the fundamental homogeneous invariants).

For D_{2n} consider the automorphism ε of $\underline{h}$ which exchanges the roots α_{2n-1} and α_{2n} and fixes all the others:

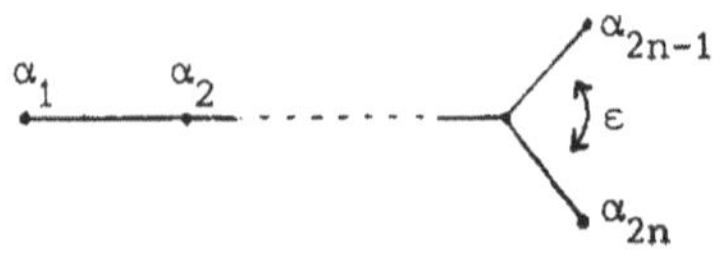

For the W-invariants $t_1,\dots,t_{2n-1},t$ of degrees $2,4,\dots,4n-2,n$ given in LIE VI § 4, n° 4.8, $t_1,\dots,t_{2n-1}$ will be invariant under ε, and $\varepsilon^*(t) = -t$. For D_4 the invariant t_3 in loc. cit. is not invariant under $\mathrm{Aut}(\Delta) \cong \mathfrak{S}_3$, however it is invariant modulo the invariants of the module $A(6)'$, since t_3 is invariant under ε. With that the statements are proven.

Lemma 2: Let σ be an automorphism of $\underline{g}$ and $\bar{\sigma}$ its image in $\mathrm{Aut}(\Delta)$. Then the determinant of the transformation σ on $\underline{g}$ is the same as the determinant of $\bar{\sigma}$ as a linear transformation on $\underline{h}/W$.

Proof: We proceed similarly as in the proof of Lemma 8.4. Let (u,t,v) be an sl_2-triplet for a regular nilpotent element $u \in \underline{g}$. The transversal slice $N = u + \underline{z}_{\underline{g}}(v)$ is stable under the action of the outer reductive centralizer $CA(u)$, which is isomorphic to $\mathrm{Aut}(\Delta)$ (cf. 7.6). From the differential characterization of regular elements (3.8, 3.14, or 7.4 Corollary 2) we obtain that the restriction of $\gamma : \underline{g} \to \underline{h}/W$ to N induces an $\mathrm{Aut}(\Delta)$-equivariant isomorphism $N \xrightarrow{\sim} \underline{h}/W$. Since the inner automorphisms of $\underline{g}$ operate by transformations of determinants 1, it therefore suffices to show

$$\det \sigma = \det(\sigma_{|N}) \quad \text{for all} \quad \sigma \in CA(u) .$$

With respect to the $CA(u)$-action we may identify N with $\underline{z}_{\underline{g}}(v)$. We regard $\underline{g}$ as an sl_2-module under the adjoint action of (u,t,v). Since the action of $CA(u)$ commutes with that of sl_2, the t-eigenspaces $N(-n)$ of weight $-n$, $n \in \mathbb{N}$, are stable under $CA(u)$ (in $\underline{z}_{\underline{g}}(v) \cong N$ only negative weights occur). Denote by $\bar{N}(-n)$ the sl_2-submodule of $\underline{g}$ generated by $N(-n)$. Then $\underline{g} = \bigoplus_{n \in \mathbb{N}} \bar{N}(-n)$, and $\bar{N}(-n)$ decomposes as a $CA(u)$-module into a direct sum of $n+1$ summands isomorphic to $N(-n)$ (cf. 7.4, or 8.4 Lemma, proof). Thus we obtain

$$\det \sigma = \prod_{n \in \mathbb{N}} \det(\sigma_{|N(-n)})^{n+1} .$$

This product runs over the $r(=\operatorname{rank}(G))$ numbers

$$d_1 - 2, \ldots\ldots, d_r - 2$$

which are all even (cf. 7.4 Proposition 2). Since $CA(u)$ is either isomorphic to $\mathbb{Z}/2\mathbb{Z}$ or to $\mathfrak{S}_3$ (cf. 7.6) we can only have $\det(\sigma_{|N(-n)}) = \pm 1$ for all n. But then

$$\det \sigma = \prod_{n \in \mathbb{N}} \det(\sigma_{|N(-n)}) = \det(\sigma_{|N}) .$$

Remark: Lemma 2 can also be checked directly. This is easy in the cases A_r and D_r, but it requires some work in the case of E_6.

We now consider $\delta : S \to \underline{h}/W$ as in the beginning of this section. According to 8.1 Lemma 2 and 8.2 we may assume δ to be of the form

$$\delta : M \oplus B \longrightarrow E \oplus B , \ \delta(m,b) = (\delta_r(m,b),b) ,$$

where M,B,E are $CA(x) \times G_m$-submodules of S and $\underline{h}/W$ with $\dim M = 3$, $\dim E = 1$, and $\dim B = r-1$ ($r = \operatorname{rank} G$). The restriction $f = \delta_{r|_M}$ is a quasihomogeneous polynomial of type $(d_r; w_r, w_{r+1}, w_{r+2})$ defining a rational double point $X_o = f^{-1}(0)$ of type Δ (cf. 8.2, 7.4 Proposition 2). Since E is the G_m-eigenspace in $\underline{h}/W$ of highest weight, the polynomial f is invariant with respect to the $CA(x)$-action on M (Lemma 1).

Lemma 3: The group $CA(x)$ operates on M by transformations of determinant 1.

Proof: We will be brief since similar arguments ocurred in the proofs of Lemma 2 and 8.4 Lemma. For the types Δ considered (i.e. A_{2r-1}, D_r, E_6) the subregular sl_2-summands in $\underline{g}$ all have odd dimension (cf. 7.4 Proposition 2), and the factor commutator group of $CA(x)$ is always $\mathbb{Z}/2\mathbb{Z}$. Since E is a trivial $CA(x)$-module we therefore obtain

$$\det\sigma = \det(\sigma_{|S}) = \det(\sigma_{|M})\cdot\det(\sigma_{|B}) = \det(\sigma_{|M})\cdot\det(\sigma_{|\underline{h}/W})$$

for all $\sigma \in CA(x)$. Because of Lemma 2 this implies $\det(\sigma_{|M}) = 1$.

Lemma 4: The action of $CA(x)$ on M is faithful.

Proof: In the case of A_{2r-1} we know already from the general part of the proof of 8.4 Lemma that the reductive centralizer $C(x) \subset CA(x)$ acts faithfully on M . Since here $CA(x)$ is the semidirect product of $C(x) \cong G_m$ with $\mathbb{Z}/2\mathbb{Z}$, we obtain the result for A_{2r-1} . In the cases D_r and E_6 the action of $CA(x)$ on the components of the Dynkin curve B_x induces the natural action of $\mathrm{Aut}(\Delta)$ on Δ . This implies the result in these cases (cf. 7.6).

Let Δ_o be an inhomogeneous irreducible Dynkin diagram such that $_h\Delta_o = \Delta$ (this is unique up to the case $\Delta = D_4$, where Δ_o may be C_3 or G_2). Define a group Γ_o of symmetries of $X_o = \delta^{-1}(\bar{0}) = f^{-1}(0)$ by

$$\Gamma_o := \begin{cases} CA(x) & \text{if } \Delta_o = C_r,\ r>3,\ F_4,\ G_2 \\ \{1,s\} & \text{if } \Delta_o = B_r,\ C_3 \text{. Here } s \text{ is a} \\ & \text{nontrivial involution } (\neq -1) \text{ in } CA(x). \end{cases}$$

(Note that for $\Delta_o = C_3$ we have $CA(x) = \mathrm{Aut}(D_4) = \mathfrak{S}_3$, and that for $\Delta_o = B_r$ we have $CA(x) = G_m \rtimes \mathbb{Z}/2\mathbb{Z}$.)

Combining now Lemma 2 and 3 with 8.4 Proposition we obtain:

Theorem: The couple (X_o,Γ_o) is a simple singularity of type Δ_o .

Now let G_o be a simple adjoint group of type Δ_o with Lie algebra $\underline{g}_o$. Let x_o be a subregular nilpotent element with sl_2-triplet (x_o,h_o,y_o) and $\delta_o : S_o \to \underline{h}_o/W_o$

the restriction of the adjoint quotient of $\underline{g}_o$ to $S_o = x_o + \underline{z}_{\underline{g}}(y_o)$.

Corollary: The Γ_o-deformation $\delta : S \to \underline{h}/W$ of X_o is Γ_o-semiuniversal, and the restriction δ^{Γ_o} of δ over the fixed point space $(\underline{h}/W)^{\Gamma_o}$ is isomorphic to δ_o .

Proof: The first statement is a consequence of Theorem 2.5. The second one follows from Theorem 2.6, Theorem 8.7 and the theorem above (note here that $\delta^{-1}(\bar{0})$ and $\delta_o^{-1}(\bar{0})$ are already isomorphic as algebraic varieties with Γ_o-action; this follows from the proof of Proposition 8.4).

Remarks: 1) It is also possible to determine the action of the reductive centralizer on the subregular singularity in the case $\Delta = A_{2r}$. By a slight modification of the arguments above one obtains the action $X,Y,Z \mapsto -X,Z,-Y$ on the singularity $X^{2r+1} + YZ = 0$ for an element $s \in CA(x)$, $s \notin C(x)$ of order 4 . The action of $\langle 1,s,s^2,s^3 \rangle$ coincides with the action of the quotient

$$D_{2r+1} / \mathbb{Z}_{2r+1} \cong \mathbb{Z}/4\,\mathbb{Z} \text{ on } \mathbb{A}^2/\mathbb{Z}_{2r+1} .$$

2) An a priori proof for the fact that $CA(x)$ operates freely on the regular points of X_o would yield a geometric proof of the theorem (by 7.6 Lemma 3 and 6.2 Theorem). The simple argument used in the proof of Theorem 8.4 for the action of $C(x)$ cannot be carried over to $CA(x)$.

3) Together with an explicit description of the $\mathrm{Aut}(\Delta)$-action on $\underline{h}/W$ the corollary above gives another proof of Proposition 8.6 for the inhomogeneous diagrams.

4) The corollary above gives an identification of $\underline{h}_o/W_o$ with $(\underline{h}/W)^{\Gamma_o}$. Such an identification can also be obtained in the following way. Since $\mathrm{Aut}(\Sigma)$ is the semi-direct product of W and $\mathrm{Aut}(\Delta)$ we may consider Γ_o as a group of automorphisms of $\underline{h}$. Let $\underline{h}_1 := \underline{h}^{\Gamma_o}$ and $W_1 := Z_W(\Gamma_o) = \{w \in W \mid w\gamma = \gamma w \text{ for all } \gamma \in \Gamma_o\}$. Then $(\underline{h}_1, W_1)$ is isomorphic to $(\underline{h}_o, W_o)$ (cf. [C] § 13, note that for $\Delta_o = B_r$ and C_r

we have isomorphic $(\underline{h}_o, W_o)$). The natural map $\underline{h}_1 \to \underline{h}/W$ induces a finite G_m-equivariant morphism $\underline{h}_1/W_1 \to (\underline{h}/W)^{\Gamma_o}$ which, by the equality of the G_m-weights, has to be an isomorphism (8.1 Lemma 3). This follows also from the fact that $Z_W(\Gamma_o)$ coincides with the stabilizer in W of $\underline{h}_1$ and that $(\underline{h}/W)^{\Gamma_o}$ is smooth.

8.9. Simultaneous Resolutions.

In 4.2 we defined when a commutative diagram of reduced varieties

$$\begin{array}{ccc} Y & \xrightarrow{\ \phi\ } & X \\ {\scriptstyle\theta}\downarrow & & \downarrow{\scriptstyle\chi} \\ V & \xrightarrow[\ \psi\ ]{} & U \end{array}$$

is a simultaneous resolution of $\chi : X \to U$. Now we will also allow X, Y, U and V to be the Henselizations of reduced varieties at closed points. Condition (iv) of 4.2, that $\phi_v : Y_v \to X_{\psi(v)}$ is a resolution for all $v \in V$, is accordingly extended to the geometric fibers over the nonclosed points of V and U . (For $k = \mathbb{C}$, one can also consider X, Y, U and V to be analytic space germs.) We consider all the deformations below to be morphisms of such Henselizations.

<u>Corollary to 8.7</u> ([Br4]): <u>Let</u> $\chi : X' \to U'$ <u>be a deformation of a rational double point</u> X_o . <u>Then</u> χ <u>possesses a simultaneous resolution.</u>

<u>Proof:</u> It follows from Theorem 8.7 and 5.3 that a simultaneous resolution for a semiuniversal deformation $X \to U$ of X_o exists:

$$\begin{array}{ccc} Y & \longrightarrow & X \\ \downarrow & & \downarrow \\ V & \longrightarrow & U \end{array}$$

Now $\chi : X' \to U'$ comes from $X \to U$ through a base change $U' \to U$ up to a

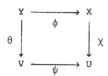

U'-isomorphism. The natural diagram

$$\begin{array}{ccc} Y \times_U U' & \longrightarrow & X \times_U U' \\ \downarrow & & \downarrow \\ V \times_U U' & \longrightarrow & U' \end{array}$$

then gives the simultaneous resolution of χ we want (the fiber products are Henselized fiber products).

Remark: For the special case when U' is smooth and one-dimensional, this result was proven directly by Brieskorn in the works [Br1] and [Br3], which suggested to Grothendieck the construction of the simultaneous resolution of the adjoint quotients $G \to T/W$ and $\underline{g} \to \underline{h}/W$ as well as the conjecture of Theorem 8.7. Generalizing the method of [Br1] and [Br3] Kas and Tjurina independently obtained the corollary above ([K] for A_r, [Tj2] in general). For generalizations and other aspects, see [Ar3], [Hui], [Pi3], [Sl2], [W] .

8.10. The Neighboring Fibers of Semiuniversal Deformations.

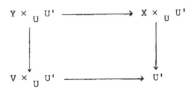

Only when, as in 8.9, we consider deformations as flat morphisms of Henselizations (or completions), we can speak of the semiuniversal deformation of the rational double point of type Δ (cf. 6.1, 2.8). However, there is a "normal form" X_o of a rational double point of type Δ with G_m-action (cf. 8.2) which possesses a G_m-equivariant semiuniversal deformation $\xi : X \to U$. With that, $X \cong \mathbb{A}^{r+2}$, $U \cong \mathbb{A}^r$, $r = \text{rank}\,\Delta$, and G_m operates on X and U with positive weights (8.6, 8.7). The properties of ξ after localization (usual or étale) are therefore essentially the same as before localization. For example, if $\xi^{-1}(u)$, $u \in U$, is a fiber of ξ , then for all neighborhoods X' of 0 in X and U' of 0 in U there exists a $u' \in U'$ with the property that $\xi^{-1}(u)$ is isomorphic to $\xi^{-1}(u')$ and all singular points of $\xi^{-1}(u')$ lie in X' (transport by G_m). In particular, the singular locus of ξ is finite over U . Therefore the following statements about the fibers of ξ translate in a suitable sense to the Henselization of ξ (here the geometric fibers over the nonclosed points have to be taken into consideration).

Definition: Let Y be a normal two-dimensional variety with isolated singular points $y_1,\dots,y_m$ and $\Delta' = \Delta_1 \cup \dots \cup \Delta_m$ a homogeneous Dynkin diagram with connected components $\Delta_1,\dots,\Delta_m$. Then Y has the singular configuration Δ', when there exists a bijection $\Delta_i \mapsto y_i$ from the components of Δ' to the singular points of Y with the property that Y has a rational double point of type Δ_i at y_i.

Corollary 1: Let $\xi : X \to U$ be a G_m-equivariant semiuniversal deformation of a simple singularity of type $\Delta = A_r, D_r, E_6, E_7, E_8$. Assume $\text{char}(k) = 0$ or $> 4\,\text{Cox}(\Delta) - 2$, and let X_u be a fiber of ξ. Then either X_u is smooth or X_u possesses the singular configuration of a subdiagram Δ' of Δ. If Δ' is a subdiagram of Δ, then there exists a fiber X_u with singular configuration Δ'.

Proof: The slice S of Theorem 8.7 fulfills the assumptions of 6.6. Now the statements follow from 8.7, 6.5 and 6.6 Corollary 2 (note the remark there for the situation for the Lie algebra $\underline{g}$).

Now let Δ be an inhomogeneous connected Dynkin diagram, ${}_h\Delta$ the associated homogeneous diagram, $AS(\Delta)$ the associated symmetry group operating on ${}_h\Delta$.

Corollary 2: Let $\xi : X \to U$ be a G_m-equivariant semiuniversal deformation of a simple singularity $(X_o, AS(\Delta))$ of type $\Delta = B_r, C_r, F_4$ or G_2. Assume $\text{char}(k) = 0$ or $> 4\,\text{Cox}(\Delta) - 2$, and let X_u be a fiber of ξ. Then either X_u is smooth or X_u possesses the singular configuration of an $AS(\Delta)$-stable subdiagram Δ' of ${}_h\Delta$, where the bijection of the components of Δ' to the singular points of X_u is equivariant with respect to $AS(\Delta)$. If Δ' is an $AS(\Delta)$-stable subdiagram of ${}_h\Delta$, then there exists a fiber X_u with singular configuration Δ'.

Proof: To the references used in the proof for Corollary 1, add 6.6 Corollary 3 .

Remarks: 1) Corollary 1 is attributed to Grothendieck by its connection with the correctness of the conjectured result of 8.7 (cf. [De2]). Up to now, many proofs by different methods have been given for Corollary 1 or its consequences (cf. [A2], [Lo], [Ly], [Pi3], [Si]; there are recent proofs by J. M. Granger and H. Laufer [Lau]).

2) It is not necessary to use the full strength of Theorem 8.7 to derive Corollaries 1 and 2. It suffices to use the fact given by 6.5 that the subregular deformation is induced from the semiuniversal deformation by a local surjective base change. The considerations in 6.6 serve (in the case of Corollary 1) only to reduce the number of singularities in the neighboring fibers to the minimal number guaranteed by the results of 6.5. Over $\mathbb{C}$ we can also use a result of Lê Dũng Tráng and Lazzeri ([Lê], [La]) which shows that the inequality $\sum_{i=1}^{m} \mu_i \leq \mu-m+1$ holds for the Milnor numbers μ_i of the singularities in a neighboring fiber of a deformation of a singularity with Milnor number μ . (For a rational double point of type Δ , Δ homogeneous, we have $\mu = \text{Rank } \Delta$).

Example: The fibers of the semiuniversal deformation of a singularity of type D_4 have the configurations D_4 , A_3 , A_2 , A_1 , $A_1 \times A_1$, $A_1 \times A_1 \times A_1$. However, in the semiuniversal deformation of a simple singularity of type G_2 only the configurations D_4 , A_1 , $A_1 \times A_1 \times A_1$ appear where the three singularities of type A_1 in the last case will be permuted by $AS(G_2) \cong \mathfrak{S}_3$. $D_4 : z^2 = X(X-\sqrt{3}\,Y)(X+\sqrt{3}\,Y)$.

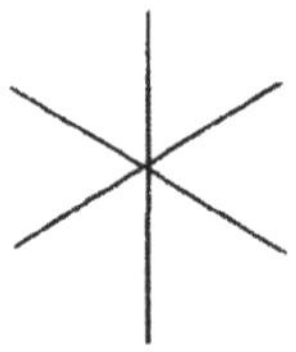

D_4

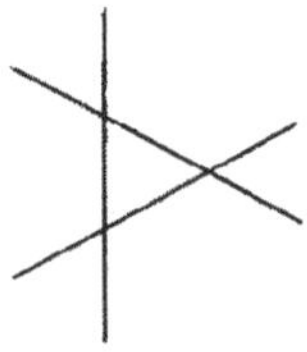

$A_1 \times A_1 \times A_1$

A_1

8.11. Other Applications.

On the basis of Theorem 8.7 we identify a G_m-equivariant semiuniversal deformation of a simple singularity of type Δ, Δ homogeneous, with the subregular deformation $\delta : S \to \underline{h}/W$ in the corresponding Lie algebra ($char(k) = 0$ or $> 4\ Cox(\Delta) - 2$).

We can stratify, first $\underline{h}$, and then $\underline{h}/W$, into a union of finitely many locally closed subsets using the type $\Delta(h)$ of the $\mathbb{Q}$-closed root system $\Sigma(h) = \{\alpha \in \Sigma \mid \alpha(h) = 0\}$. For $\underline{h}/W$ this stratification is equivalent to that induced by the (singular) configuration type of the fibers of δ (6.5). In particular, the singular fibers of δ lie exactly over the discriminant $D_{\underline{h}/W}$, which is the image of the union $\underline{h}_{\Sigma}$ of the root hyperplanes $\underline{h}_{\alpha} = \{h \in \underline{h} \mid \alpha(h) = 0\}$ under the quotient map $\underline{h} \to \underline{h}/W$. Since the roots of a homogeneous root system are conjugate under the Weyl group, $D_{\underline{h}/W}$ is already the image of a single root hyperplane $\underline{h}_{\alpha}$, and in particular is irreducible.

For a topological investigation of the complement of the discriminant of δ made possible by the description above, see [Br5].

The irreducibility of the discriminant is a property of the semiuniversal deformation for any hypersurface with isolated singularities. Other such properties which can be derived for the simple singularities by the description of 8.7 are, for example, the theorem of Lê and Lazzeri cited in 8.10 or the openness of the versality ([Te] III) in the base of a semiuniversal deformation (here note the remark to the proof of Lemma 3.10). To close we mention an application of 8.7 to questions about the Lie algebra.

Corollary: Let $\underline{g}$ be a Lie algebra of a reductive group G, whose Dynkin diagram (i.e. that of (G,G)) is homogeneous. If $char(k) = 0$ or $char(k) > 4\ Cox(G_i) - 2$ for every simple normal subgroup of G, then the subregular elements of $\underline{g}$ form a locally-closed smooth subvariety of codimension 3 in $\underline{g}$. If $char(k) = 0$, the corresponding statement also holds for the subregular elements in G.

Proof: First let G be simple. A simple calculation shows that the critical locus of a semiuniversal deformation of a two-dimensional isolated hypersurface singularity is smooth and has codimension 3 in the total space. Therefore the subregular elements of a transverse slice as in 8.7 form a smooth subvariety of S. Because the morphism $G \times S \to \underline{g}$, $(g,s) \mapsto Ad(g)s$, is smooth, the subregular orbits meeting S form a locally closed smooth subvariety of $\underline{g}$. By 6.5, S meets all subregular orbits over a neighborhood of $\overline{0} \in \underline{h}/W$, and therefore all subregular orbits in $\underline{g}$, because of the quasihomogeneous structure of S (cf. 7.4) and the linearity of the G-action on $\underline{g}$. (The statement about the codimension also follows from 4.6.) If G is reductive, then the statement about G reduces to a statement about the simple normal subgroups of G by an analysis similar to that given in 5.4.

We now consider the subregular elements of G (for $char(k) = 0$). By the Comparison Theorem 3.15 we first have the smoothness of the subregular set over a neighborhood of $\overline{e}$ in T/W. The subregular orbits "far away" from $e \in G$ can no longer be reached by a G_m-action (scalars in $\underline{g}$). So instead of that, we use the Slice Theorem of Luna ([Lu]). Every closed orbit of G is semisimple and meets T. For all $t \in T$, $Z_G(t)^o$ is locally a transverse slice to the G-orbit of t, and the G-morphism

$$G \times^{Z_G(t)} Z_G(t)^o \to G \ , \ g * z \mapsto {}^g(tz) \ ,$$

is étale in a neighborhood of $G \times^{Z_G(t)} V(t)$ (where $V(t) \subset Z_G(t)^o$ is the unipotent variety). The validity of our global statement about the subregular elements in G now follows from the validity of the local statements for the reductive subgroups $Z_G(t)^o$ (the root system of $Z_G(t)^o$ is homogeneous), since the étale open sets of G given by the morphisms above cover G (cf. 3.10).

Remark: This Corollary was already formulated in [Br4].

For groups with inhomogeneous Dynkin diagram the Corollary is false. As an example, consider the case G_2 . We will calculate the intersection of the subregular slice with the subregular elements, or equivalently the critical locus of a semiuniversal deformation of a simple singularity $(X_o, \mathfrak{S}_3)$ of type G_2 . We let $X_o = \{(x,y,z) \in \mathbb{A}^3 \mid z^2 = x^3 - 3xy^2\}$. Then $\mathfrak{S}_3$ acts on the z-axis by sign, and on the (x,y)-plane it acts orthogonally with respect to $x^2 + y^2$ by permutation of the three linear factors of $x(x^2 - 3y^2)$. A semiuniversal deformation of $(X_o, \mathfrak{S}_3)$ is defined by $\phi : \mathbb{A}^4 \to \mathbb{A}^2$, $(x,y,z,w) \mapsto (x^3 - 3xy^2 - z^2 + 3w\,(x^2 + y^2), w)$. The critical locus of ϕ lies in the subspace $\{z = 0\}$ of $\mathbb{A}^4$ and there decomposes into 4 components:

$$C_o = \{y = x = 0\} , \quad C_{1,2} = \{x = w , y = \pm \sqrt{3}\, x\} ,$$

$$C_3 = \{y = 0 , x = -2w\}$$

The projection of C_1 , C_2 , C_3 to the (x,y)-plane gives the three lines of the equation $y^3 - 3yx^2 = 0$. Therefore C_1 , C_2 and C_3 are permuted by $\mathfrak{S}_3$. The action of $\mathfrak{S}_3$ on the w-axis, i.e. C_o , is trivial.

The image under ϕ of C_o is the w-axis in $\mathbb{A}^2$, and the image of C_i , $i = 1,2,3$, is the parabola $\{(4w^3, w)\}$. The discriminant therefore has two components where the neighboring fibers have a singularity of type A_1 over one and 3 singularities of type A_1 over the other which will be permuted by $\mathfrak{S}_3$ (cf. the example in 8.10).

By 4.6 and 4.7 the set of irregular elements of a simple Lie algebra $\underline{g}$ decomposes into either one or two components depending on whether the Dynkin diagram of $\underline{g}$ is homogeneous or inhomogeneous. The subregular orbits are open and dense in these components. In a similar way as above one may calculate the structure of the subregular set for Δ of type B_r , C_r , F_4 . Here one always obtains two smooth components intersecting transversely. As the above example shows, the singularities of the subregular set may be worse, i.e. a component itself may be singular (in the example, C_1 , C_2 , C_3 belong to the same irreducible global component. This example was independently found by Dale Peterson [Pe]).

Appendix I: Forms of Simple Singularities and Simple Algebraic Groups

Let (X,x) be a rational double point of type $\Delta = A_r$, D_r , E_r over an algebraically closed field of good characteristic. In 6.4 (and 8.3) we have shown how (X,x) may be realized as the "generic" singularity of the unipotent (resp. nilpotent) variety of a corresponding almost simple group G (resp. its Lie algebra). Here we will extend this result to not necessarily algebraically closed fields. We will only state the main results using freely the concepts of the relative theory of semisimple groups ([Bo-Ti], [Ti]). Details are left to a future work. To simplify the presentation we assume the base field k to be perfect and of zero or sufficiently high characteristic.

In the following discussion we list the possible k-forms of (X,x) (up to Henselisation) together with that k-form of G whose unipotent variety realizes the singularity in question along its subregular orbit. Only such forms of G occur which possess k-rational subregular elements. These forms can be classified by the "index" attached to them (cf. [Ti] 2.3). More precisely, one can show that a unipotent class of G possesses a k-rational element if and only if its valuated Dynkin diagram is compatible with the index of G , i.e. if the valuation is symmetric with respect to the Galois-action on the Dynkin diagram Δ and if the values are zero at the anisotropic roots (Δ_o in loc. cit.).

The classification of the k-forms of the rational double points was essentially done by Lipman ([Li] § 24) who associates to them a Dynkin diagram of homogeneous or inhomogeneous type. All diagrams A_r , B_r ,..., G_2 actually occur. Yet, the correspondence leaves some ambiguities and cannot be carried over to the group-theoretic interpretation. Therefore we will replace Lipman's diagram by the index of the corresponding group. This invariant leaves no ambiguities and determines the divisor class group H in a natural way, i.e. $H = L^*/L$ where L^* (resp. L) is the weight (resp. root) lattice of the relative root system which can be derived from the index ([Ti] 2.5).

1) Forms of A_{2n-1}.

a) The split form is given by

$$X^{2n} + Y^2 - Z^2 = 0 .$$

The index of the corresponding group $(Sl_{2n}(k))$ is

The relative root system is of type A_{2n-1} and the divisor class group $H = L^*/L$ is $\mathbb{Z}/2n.\mathbb{Z}$ (Lipman type A_{2n-1}).

b) The quasi-split forms are given by

$$X^{2n} + aY^2 - Z^2 = 0 , n \geq 2 ,$$

where $a \in k$ is not a square in k. The index of the corresponding groups $(SU_{2n}(K,h)$, where K is the quadratic extension of k determined by a and h is a nondegenerate hermitian form of maximal Witt index n) is

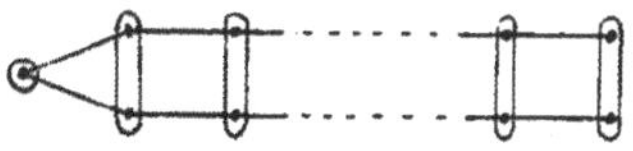

The relative root system is C_n , and $H = \mathbb{Z}/2\,\mathbb{Z}$. (Lipman type B_n).

c) The "weakly anisotropic" forms are given by

$$X^{2n} + aY^2 - bZ^2 = 0$$

where the quadratic form $Q = X^2 + aY^2 - bZ^2$, $a,b \in k$, has no nontrivial zero over k . The index of the corresponding groups ($SU_{2n}(K,h)$ where $K = k(\sqrt{ab})$ and h is a hermitian form of Witt index $n - 1$ and discriminant $-a \bmod N_{K/k}(K^*)$) is

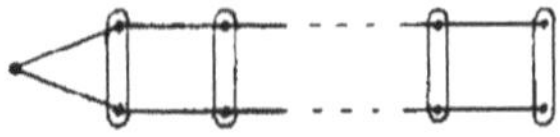

The relative root system is BC_{n-1} , and H is trivial. (Lipman type B_n).

2) Forms of A_{2n}

a) The split form is defined by

$$X^{2n+1} + Y^2 - Z^2 = 0 .$$

The index of the corresponding group $(SL_{2n+1}(k))$ is

with relative root system of type A_{2n} . The divisor class group is $H = \mathbb{Z}/(2n+1)\mathbb{Z}$. (Lipman type A_{2n}).

b) The quasi-split forms are given by

$$X^{2n+1} + Y^2 - aZ^2 = 0$$

where $a \in k$ is not a square. The index of the corresponding groups ($SU_{2n+1}(K,h)$ where K is the quadratic extension defined by a and where h is a nondegenerate hermitian form of maximal Witt index n) is

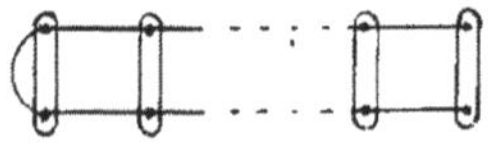

with relative root system of type BC_n . We have $H = 1$. (Lipman type B_n).

3) Forms of D_{2n} , $n \geq 2$

a) Split form

$$X^{2n-1} - XY^2 + Z^2 = 0$$

Group: $SO_{4n}(q)$, q a nondegenerate quadratic form of maximal Witt index $2n$

Index:

Relative root system: D_{2n}

$$H = (\mathbb{Z}/2\mathbb{Z}) \times (\mathbb{Z}/2\mathbb{Z}) . \quad \text{(Lipman type } D_{2n} \text{)}.$$

b) Quasi-split forms

$$X^{2n-1} - aXY^2 + Z^2 = 0 , \; a \in k\backslash k^2 .$$

Group: $SO_{4n}(q)$, q a nondegenerate quadratic form of Witt index $2n-1$ and discriminant a .

Index:

Relative root system: B_{2n-1}

$$H = \mathbb{Z}/2\mathbb{Z} . \quad \text{(Lipman type } C_{2n-1} \text{)}$$

c) Trialitary quasi-split forms of D_4

$$Q(X,Y) + Z^2 = 0$$

where Q is a nondegenerate cubic form with no nontrivial zeroes over k .

Group: Quasi-split trialitary form of type D_4 (${}^3D^2_{4,2}$ or ${}^6D^2_{4,2}$ in [Ti] p. 58).

Index:

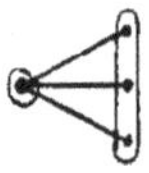

Relative root system: G_2

$$H = 1 \qquad \text{(Lipman type } G_2 \text{)}$$

4) Forms of D_{2n+1} , $n \geq 2$

a) Split form

$$X^{2n} + XY^2 - Z^2 = 0$$

Group: $SO_{4n+2}(q)$, q a nondegenerate quadratic form of maximal Witt index $2n+1$.

Index:

Relative root system: D_{2n+1}

$$H = \mathbb{Z}/4\mathbb{Z} \qquad \text{(Lipman type } D_{2n+1} \text{)}$$

b) Quasi-split forms

$$X^{2n} + XY^2 - aZ^2 \quad , \quad a \in k \setminus k^2$$

Group: $SO_{4n+2}(q)$, q a nondegenerate quadratic form of Witt index 2n and discriminant a .

Index:

Relative root system: B_{2n}

$$H = \mathbb{Z}/2\,\mathbb{Z} \,. \qquad \text{(Lipman type } C_{2n}\text{)}$$

5) Forms of E_6

a) Split form

$$X^4 + Y^3 - Z^2 = 0$$

Group: Chevalley group of type E_6

Index:

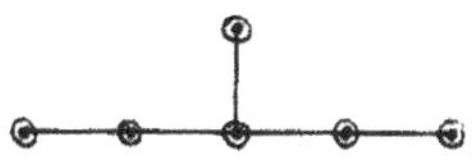

Relative root system: E_6

$$H = \mathbb{Z}/3\,\mathbb{Z} \,. \qquad \text{(Lipman type } E_6\text{)}$$

b) Quasi-split forms

$$X^4 + Y^3 - aZ^2 = 0 \quad , \quad a \in k \setminus k^2 \,.$$

Group: Quasi-split group of type E_6 with respect to $k(\sqrt{a})$.

Index:

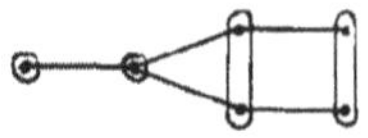

Relative root system: F_4

$$H = 1 \, . \qquad \text{(Lipman type } F_4 \text{)}$$

6) Forms of E_7

Split form

$$X^3Y + Y^3 + Z^2 = 0$$

Group: Chevalley group of type E_7

Index:

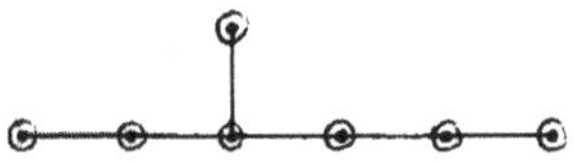

Relative root system: E_7

$$H = \mathbb{Z}/2\,\mathbb{Z} \, . \qquad \text{(Lipman type } E_7 \text{)}$$

7) Forms of E_8

Split form

$$X^5 + Y^3 + Z^2 = 0$$

Group: Chevalley group of type E_8

Index:

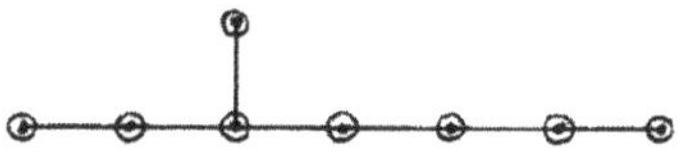

Relative root system: E_8

$$H = 1 . \quad \text{(Lipman type } E_8 \text{)}$$

One may ask what forms of singularities occur in forms of groups of type B_r , C_r , F_4 , G_2 . Here the situation becomes more complicated:

In a Chevalley group G of inhomogeneous type the subregular orbit decomposes into several orbits under the group $G(k)$ of k-rational points of G . Accordingly the split and all quasi-split k-forms of the rational double point of type ${}_h\Delta$ are realized. Moreover, the associated symmetry group is not always conserved. Hence these singularities are not k-forms of a simple singularity of inhomogeneous type Δ . This seems to be natural since the interpretation of the k-forms as quotient singularities also breaks down.

The only form of a group of inhomogeneous type which is not a Chevalley group and yet possesses a subregular unipotent k-rational element is (up to isogeny) $SO_{2r+1}(q)$, where q is a quadratic form of Witt index $r - 1$ and anisotropic part $q_o = X^2 + aY^2 - bZ^2$. The index is

and the subregular singularity is the "weakly anisotropic" form of A_{2r-1} with $Q = q_o$ (cf. 1) c), above).

Theorem 8.7 stating the semiuniversality of the subregular deformations remains valid for A_r , D_r , E_r without any restriction. In cases B_r , C_r , F_4 , G_2 one has to restrict to the forms with full symmetry.

Appendix II: A Semiuniversality Property of Adjoint Quotients

Let G be a linearly reductive group and X_o a G-complete intersection defined by a flat G-equivariant morphism $f : V \to W$ of finite-dimensional linear G-spaces (cf. 2.5). If X_o has isolated singularities then a semiuniversal deformation of the couple (X_o, G) exists by 2.6 Corollary. The proof of this fact was easily derived from 2.5 Theorem. One can show that the condition on X_o to have isolated singularities can be relaxed if the group G is positive-dimensional. More precisely: Let $T^1(f)$ denote the cokernel of the G-homomorphism

$$Tf \; : \; k[V] \otimes V \to k[X_o] \otimes W$$

induced by the differential of f (cf. proof of 2.5 Theorem). Then one can prove (details will appear elsewhere):

Theorem: A semiuniversal deformation of (X_o, G) exists exactly when the G-invariant part $T^1(f)^G$ of $T^1(f)$ has finite dimension over the base field k.

If $T^1(f)^G$ is finite-dimensional then a semiuniversal deformation of (X_o, G) can be constructed in a similar way as was done in the proof of 2.5 Theorem. As a corollary one obtains:

Corollary: Let V be a linear G-space such that the quotient morphism $\pi : V \to V/G$ is flat. Then π is a versal deformation of (X_o, G) where $X_o = \pi^{-1}(\pi(0))$ is equipped with the induced G-action. Moreover, π is semiuniversal exactly when $V^G = \{0\}$.

Remark: The flatness of π implies that V/G is isomorphic to an affine space.

It follows from the corollary that the adjoint quotient $\gamma : \underline{g} \to \underline{h}/W$ for a semisimple Lie algebra $\underline{g}$ (over a field k of characteristic 0) is a semiuniversal deformation of the nilpotent variety $N(\underline{g}) = \gamma^{-1}(\gamma(0))$ equipped with the natural G-action.

Question: Can one use this result to prove Theorem 8.7? More generally: Does the semiuniversality of γ imply the versality of γ restricted to transversal slices S which are equipped with natural actions by (reductive) centralizer subgroups? By Luna's Slice Theorem ([Lu]) the answer is "yes" for slices to closed orbits.

Linear representations V of simple groups G whose quotient $\pi : V \to V/G$ is flat have recently been classified by Popov and Schwarz ([P], [S]). It would be interesting to study the generic singularities of the corresponding "nilpotent" varieties $\pi^{-1}(\pi(0))$ as well as their deformations.

Appendix III: Dynkin Diagrams and Representations of Finite Subgroups of SL_2

Let F be a finite subgroup of SL_2 (for simplicity, say over $\mathbb{C}$). In 6.1 we have associated to F a homogeneous Dynkin diagram $\Delta(F)$ by looking at the minimal resolution of the quotient singularity $\mathbb{C}^2/F$. Recently a purely group-theoretic definition of the correspondence

$$F \longmapsto \Delta(F)$$

was found by John McKay (Montreal): Let $R_o, R_1, \ldots, R_r$ (resp. N) denote the equivalence classes of the irreducible representations of F (resp. of a fixed natural representation $F \hookrightarrow SL_2$, like that of 6.1). Define the $(r+1)\times(r+1)$-matrix $A = ((a_{ij}))$ by the decomposition formula

$$N \otimes R_i = \bigoplus_{j=o}^{r} a_{ji} R_j ,$$

where a_{ji} denotes the multiplicity of R_j in $N \otimes R_i$. Denote the identity matrix by I.

Then

$$C = 2I - A$$

is the Cartan matrix of the extended Dynkin diagram $\tilde{\Delta}(F)$ *of* $\Delta(F)$.

The classes of irreducible representations of F thus correspond bijectively to the vertices of the extended Dynkin diagram $\tilde{\Delta}(F)$. One may choose the additional point of $\tilde{\Delta}(F)$ to correspond to the one-dimensional trivial representation. The dimensions of the representations corresponding to points of $\Delta(F) \subset \tilde{\Delta}(F)$ then coincide with the coefficients of the highest root in the root system of $\Delta(F)$. This last statement is a particular case of the interpretation of the columns of the character table of F as eigenvectors of A (and C), which follows from the equation $N \otimes R_i = \bigoplus a_{ji} R_j$.

Example: Let $F = \mathbb{I}$ be the binary icosahedral group. Then $\tilde{\Delta}(F)$ is the extended Dynkin diagram of E_8 :

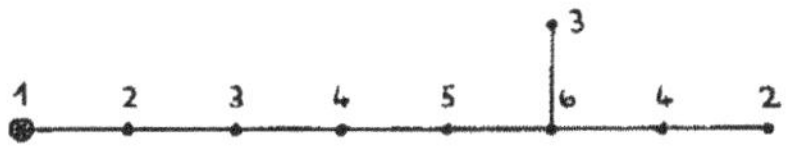

The numbers attached are the dimensions of the representations.

Now we will describe a group-theoretic interpretation of the inhomogeneous Dynkin diagrams, more precisely, of the Dynkin diagrams appearing in the theory of reduced affine root systems (Kac, MacDonald, Moody, Bruhat-Tits).

In 6.2 we have related certain couples of groups $F \triangleleft F' \subset SL_2$ with inhomogeneous Dynkin diagrams $\Delta(F,F')$:

$\Delta(F,F')$	F	F'
B_r	$\mathbb{Z}_{2r}$	$\mathbb{D}_r$
C_r	$\mathbb{D}_{r-1}$	$\mathbb{D}_{2(r-1)}$
F_4	$\mathbb{T}$	$\mathbb{O}$
G_2	$\mathbb{D}_2$	$\mathbb{T}$

(In the case G_2 we have replaced the group $\mathbb{O}$ by the smaller group $\mathbb{T}$. This simplifies the following description. Moreover, the theorems in 8.4 and 8.7 concerning G_2 remain valid when reformulated accordingly.)

Now fix a couple $F \triangleleft F'$ as above. By restriction, the irreducible representations of F' may be regarded as representations of F . Let $S_1,\ldots,S_n$ denote the equivalence classes (with respect to F) of these representations and let N be the fixed natural representation of F which may be considered as the restriction of the fixed natural representation of F' . Then the following decomposition formula makes sense

$$N \otimes S_i = \bigoplus b_{ji} S_j$$

and defines a uniquely determined $n \times n$-matrix $B = ((b_{ij}))$. One verifies that

$$C = 2I - B$$

is the Cartan matrix of the extended Dynkin diagram $\widetilde{\Delta^*(F,F')}$ of the dual of $\Delta(F,F')$ (note $B_r^* = C_r$, $C_r^* = B_r$, $F_4^* = F_4$, $G_2^* = G_2$).

Similarly we may look at the F'-equivalence classes $Q_1,\dots,Q_m$ of representations of F' which are induced from irreducible representations of F . Then $m = n$ and, with respect to a conveniently chosen ordering of the Q_i , the following decomposition formula holds

$$N \otimes Q_i = \bigoplus b_{ij} Q_j ,$$

i.e. the decomposition of the induced representations is described by ${}^tB$. The matrix $2I - {}^tB$ is the Cartan matrix of the dual of $\widetilde{\Delta^*(F,F')}$. Thus the restricted or induced representations correspond bijectively to the vertices of an inhomogeneous affine Dynkin diagram.

1) $\mathbb{Z}_{2r} \lhd \mathbb{D}_r$

$\widetilde{\Delta(F')} = \tilde{D}_{r+2}$

$\widetilde{\Delta^*(F,F')} = \tilde{C}_r$

$(\tilde{C}_r)^*$

$\widetilde{\Delta(F)} = \tilde{A}_{2r-1}$

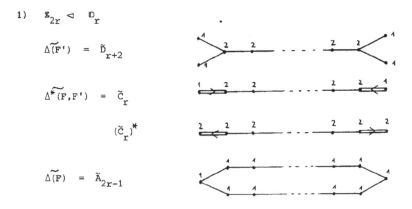

2) $\mathbb{D}_{r-1} \lhd \mathbb{D}_{2(r-1)}$

$\widetilde{\Delta(F')} = \tilde{D}_{2r}$

$\Delta^*\widetilde{(F,F')} = \tilde{B}_r$

$(\tilde{B}_r)^*$

$\widetilde{\Delta(F)} = \tilde{D}_{r+1}$

3) $\mathbb{T} \lhd \mathbb{O}$

$\widetilde{\Delta(F')} = \tilde{E}_7$

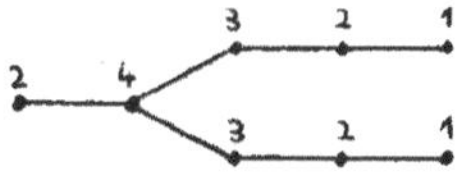

$\Delta^*\widetilde{(F,F')} = \tilde{F}_4$

$(\tilde{F}_4)^*$

$\widetilde{\Delta(F)} = \tilde{E}_6$

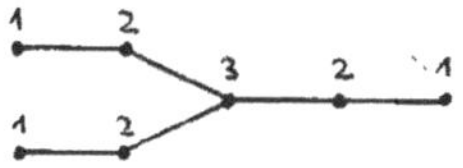

4) $\mathbb{D}_2 \lhd \mathbb{T}$

$\widetilde{\Delta(F')} = \tilde{E}_6$

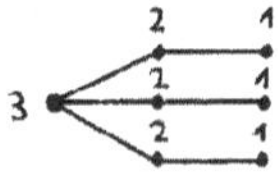

$\Delta^*\widetilde{(F,F')} = \tilde{G}_2$

$(\tilde{G}_2)^*$

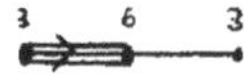

$\widetilde{\Delta(F)} = \tilde{D}_4$

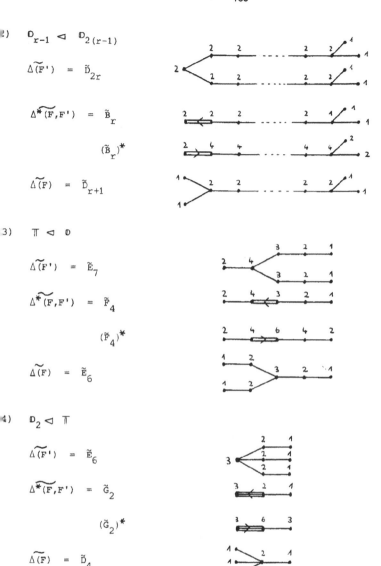

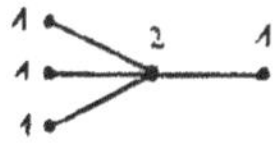

The numbers attached to the vertices of the Dynkin diagrams above are the dimensions of the corresponding representations. They coincide (in case of $(\Delta^*\widetilde{(F,F')})^*$: up to the factor $[F' : F]$) with the coefficients attached to them in the theory of affine root systems (cf. for example [MD]). Again, this is related to an eigenvector interpretation of the columns of the restricted and induced character tables (analogous to the homogeneous case).

Let us note the following property which appears above in the description of the induction. The quotient group F'/F acts in a natural way on the set of equivalence classes of representations of F . Since N is kept invariant we obtain an action of F'/F on the extended Dynkin diagram $\widetilde{\Delta(F)}$ and on the subdiagram $\Delta(F)$. The last action coincides with the associated action as defined in 6.2.

One may define the correspondence

$$F \longmapsto \Delta(F)$$

also by going the inverse direction. Starting with the Cartan matrix $C = ((c_{ij}))$ of a homogeneous Dynkin diagram Δ of type A_r , D_r , E_r one may obtain the group F with $\Delta(F) = \Delta$ by defining F to be the group generated by r elements $e_1,\ldots,e_r$, $r = \operatorname{rank}(\Delta)$, subject to r relations

$$e_1^{c_{i1}} \cdot e_2^{c_{i2}} \cdot \ldots \cdot e_r^{c_{ir}} = 1 \quad , \quad i = 1,\ldots,r \; .$$

This is easily seen either by a simple reduction to other well known sets of generators and relations (cf. [C-M] 6.5) or by Mumford's calculation of the local fundamental group of rational surface singularities (cf. [M1], [Br3]).

To conclude this appendix and with it the preceding chapters we advise the reader to read (once more) the closing remarks of Brieskorn in his report at Nice (cf. [Br4], [Kl] Vorrede).

Bibliography

[A 1] Arnol'd, V. I.: On matrices depending on parameters, Russian Math. Surveys 26, 2; 29 - 43 (1971)

[A 2] Arnol'd, V. I.: Normal forms for functions near degenerate critical points, the Weylgroups of A_k , D_k , E_k and Lagrangian singularities, Functional Anal. Appl. 6, 254 - 272 (1972)

[A 3] Arnol'd, V. I.: Normal forms of functions in a neighborhood of a degenerate critical point, Russian Math. Surveys 29, 2; 11 - 50 (1974)

[A 4] Arnol'd, V. I.: Critical points of functions on a manifold with boundary, the simple Lie groups B_k , C_k , F_4 and singularities of evolutes, Russian Math. Surveys 33, 5, 99 - 116 (1978)

[Ar 1] Artin, M.: Some numerical criteria for contractability of curves on algebraic surfaces, Amer. J. Math. 84, 485 - 496 (1962)

[Ar 2] Artin, M.: On isolated rational singularities of surfaces, Amer. J. Math. 88, 129 - 136 (1966)

[Ar 3] Artin, M.: An algebraic construction of Brieskorn's resolutions, J. of Algebra 29, 330 - 348 (1974)

[Ar 4] Artin, M.: Lectures on deformations of singularities, Tata Institute, Bombay, 1976

[Ar 5] Artin, M.: Coverings of the rational double points in characteristic p , in: Complex Analysis and Algebraic Geometry, ed. W. L. Baily, jr. and T. Shioda, Iwanami Shoten, Publ., Cambridge Univ. Press, 1977

[Bo] Borel, A.: Linear Algebraic Groups, Benjamin, New York, 1969

[Bo-Ti] Borel, A., Tits, J.: Groupes Réductifs, Publ. Math. I.H.E.S. 27, 55 - 152 (1965)

[B-Kr] Borho, W., Kraft, H.: Über Bahnen und deren Deformationen bei linearen Aktionen reduktiver Gruppen, Commentarii Math. Helvetici 54, 61 - 104 (1979)

[Br 1] Brieskorn, E.: Über die Auflösung gewisser Singularitäten von holomorphen Abbildungen, Math. Ann. 166, 76 - 102 (1966)

[Br 2] Brieskorn, E.: Rationale Singularitäten komplexer Flächen, Inventiones math. 4, 336 - 358 (1968)

[Br 3] Brieskorn, E.: Die Auflösung der rationalen Singularitäten holomorpher Abbildungen, Math. Ann. 178, 255 - 270 (1968)

[Br 4] Brieskorn, E.: Singular elements of semisimple algebraic groups, in: Actes Congrès Intern. Math. 1970, t. 2, 279 - 284

[Br 5] Brieskorn, E.: Die Fundamentalgruppe des Raumes der regulären Orbits einer endlichen komplexen Spiegelungsgruppe, Inventiones math. 12, 57 - 61 (1971)

[C] Carter, R. W.: Simple groups of Lie type, Wiley and Sons, London-New York, 1972

[C-M] Coxeter, H. S. M., Moser, W. O. J.: Generators and Relations for Discrete Groups, 3rd edition, Springer, Berlin-Heidelberg-New York 1975

[De 1] Demazure, M.: Invariants symétriques entiers des groupes de Weyl et torsion, Inventiones math. 21, 287 - 301 (1973)

[De 2] Demazure, M.: Classification des germes à point critique isolé et à nombre de modules 0 ou 1, in: Séminaire Bourbaki n° 443, Lect. Notes in Math. 431, Springer, Berlin-Heidelberg-New York, 1975

[De-Ga] Demazure, M., Gabriel, P.: Groupes algébriques I, Masson-North Holland, Paris-Amsterdam, 1970

[DV] Du Val, P.: On isolated singularities which do not affect the conditions of adjunction, Part I, Proc. Cambridge Phil. Soc. 30, 453 - 465 (1934)

[Dy] Dynkin, E. B.: Semisimple subalgebras of semisimple Lie algebras, A.M.S. Translations, Ser. 2, 6, 111 - 245 (1957)

[EGA] Grothendieck, A., Dieudonné, J. A.: Eléments de géométrie algébrique, I, Springer, Berlin-Heidelberg-New York, 1971; II, III, IV, Publ. Math. I.H.E.S. 8, 11, 17, 20, 24, 28, 32, 1961-67

[El] Elkington, G. B.: Centralizers of unipotent elements in semisimple algebraic groups, J. of Algebra 23, 137 - 163 (1972)

[Es] Esnault, H.: Singularités rationelles et groupes algébriques, Thèse de 3ème cycle, Paris VII, 1976

[Ga] Gabriel, P.: Unzerlegbare Darstellungen I, Manuscripta math. 6, 71 - 103 (1972)

[G] Grothendieck, A.: Sur quelques propriétés fondamentales en théorie des intersections, in: Séminaire C. Chevalley: Anneaux de Chow et applications, Secréteriat mathématique, Paris, 1958

[H-C] Harish-Chandra: Invariant distributions on Lie algebras, Amer. J. Math. 86, 271 - 309 (1964)

[He] Hesselink, W.: Singularities in the nilpotent scheme of a classical group, Proefschrift Utrecht, 1975, for a published version see Transact. A.M.S. 222, 1 - 32 (1976)

[Hi] Hironaka, H.: Resolution of singularities of an algebraic variety over a field of characteristic zero, Ann. of Math. 79, 109 - 326 (1964)

[Hui] Huikeshoven, F.: On the versal resolutions of deformations of rational double points, Inventiones math. 20, 15 - 33 (1973)

[Hu] Humphreys, J. E.: Linear Algebraic Groups, Springer, Berlin-Heidelberg-New York, 1975

[J] Jacobson, N.: A note on Three-Dimensional Simple Lie algebras, J. of Math. and Mech. 7, 5, 823 - 831 (1958)

[Jä] Jänich, K.: Differenzierbare G-Mannigfaltigkeiten, Lecture Notes in Math. 59, Springer, Berlin-Heidelberg-New York, 1968

[K] Kas, A.: On the resolution of certain holomorphic mappings, Amer. J. Math. 90, 789 - 804 (1968)

[K-S] Kas, A,. Schlessinger, M.: On the versal deformation of a complex space with an isolated singularity, Math. Ann. 196, 23 - 29 (1972)

[Kl] Klein, F.: Vorlesungen über das Ikosaeder und die Auflösung der Gleichungen vom fünften Grade, Teubner, Leipzig 1884

[Ko 1] Kostant, B.: The principal three-dimensional subgroup and the Betti numbers of a complex simple Lie group, Amer. J. Math. 81, 973 - 1032 (1959)

[Ko 2] Kostant, B.: Lie group representations on polynomial rings, Amer. J. Math. 85, 327 - 404 (1963)

[La] Lazzeri, F.: A theorem on the monodromy of isolated singularities, Asterisque 7 et 8, 269 - 275 (1973)

[Lau] Laufer, H. B.: Ambient deformations for exceptional sets in two-manifolds, Inventiones math. 55, 1 - 36 (1979)

[Lê] Lê Dung Tràng: Une application d'un theoreme d'A'Campo a l'equisingularité, Ecole Polytechnique, Centre des Math., Preprint N° A 113.0273, (1973)

[LIE] Bourbaki, N.: Groupes et algèbres de Lie, I - VIII, Hermann, Paris, 1971, 1972, 1968, 1975

[Li] Lipman, J.: Rational singularities, with applications to algebraic surfaces and unique factorization, Publ. Math. I.H.E.S. 36, 195 - 279 (1969)

[Lo] Looijenga, E. J.: A period mapping for certain semi-universal deformations, Compositio Math. 30, 299 - 316 (1975)

[Lu] Luna, D.: Slices étales, Bull. Soc. Math. France, Mémoire 33, 81 - 105 (1973)

[Lus] Lusztig, G.: On the finiteness of the number of unipotent classes, Inventiones math. 34, 201 - 213 (1976)

[Ly] Lyashko, O. V.: Decomposition of simple singularities of functions, Functional Anal. Appl. 10, 2, 122 - 128 (1976)

[MD] Macdonald, I. G.: Affine root systems and Dedekind's η-function, Inventiones math. 15, 91 - 143 (1972)

[M 1] Mumford, D.: The topology of normal singularities of an algebraic surface and a criterion for simplicity, Publ. Math. I.H.E.S. 9, 5 - 22 (1961)

[M 2] Mumford, D.: Geometric invariant theory, Springer, Berlin 1965

[Na] Nagata, M.: Complete reducibility of rational representation of a matric group, J. Math. Kyoto Univ. 1, 87 - 99 (1961)

[P] Popov, V. L.: Representations with a free module of covariants, Functional Anal. Appl. 10, 242 - 244 (1977)

[Pe] Peterson, D.: Geometry of the adjoint representation of a complex semi-simple Lie algebra, Thesis, Harvard-University, Cambridge Mass., 1978

[Pi 1] Pinkham, H. C.: Deformations of algebraic varieties with G_m-action, Asterisque 20 (1974)

[Pi 2] Pinkham, H. C.: Deformations of normal surface singularities with C -action, Math. Ann. 232, 65 - 84 (1978)

[Pi 3] Pinkham, H. C.: Séminaire sur les singularités des surfaces, exposés du 12.10.76, 26.10.76, 4.1.77, 18.1.77, Ecole Polytechnique, Centre des Math., année 1976-77

[Po] Poénaru, V.: Singularités C^∞ en présence de symétrie, Lecture Notes in Mathematics No 510, Springer, Berlin-Heidelberg-New York 1976

[Ri 1] Richardson, R. W.: Conjugacy classes in Lie algebras and algebraic groups, Ann. of Math. 86, 1 - 15 (1967)

[Ri 2] Richardson, R. W.: Conjugacy classes in parabolic subgroups of semisimple algebraic groups, Bull. London Math. Soc. 6, 21 - 24 (1974)

[Rim 1] Rim, D. S.: Formal deformation theory, in SGA 7, I, Lecture Notes in Math. No 288, Springer, Berlin-Heidelberg-New York, 1972

[Rim 2] Rim, D. S.: Equivariant G-structure on versal deformations, Transact. A.M.S. 257, 217 - 226 (1980)

[Ro] Rosenlicht, M.: On quotient varieties and the affine embedding of certain homogeneous spaces, Transact. A.M.S. 101, 211 - 223 (1961)

[Schl] Schlessinger, M.: Functors of Artin rings, Transact. A.M.S. 130, 205 - 222 (1968)

[S] Schwarz, G. W.: Representations of simple Lie groups with a free module of covariants, Inventiones math. 50, 1 - 12 (1978)

[Se] Serre, J. P.: Expaces fibrês algébriques, in: Séminaire C. Chevalley: Anneaux de Chow et applications, Secrétariat mathématique, Paris, 1958

[Sh] Shoji, T.: Conjugacy classes of Chevalley groups of type F_4 over finite fields of characteristic $p \neq 2$, Journ. Fac. Sci. Tokyo Univ. 21, 1 - 17 (1974)

[Si] Siersma, D.: Classification and deformation of singularities, Proefschrift Amsterdam, 1974

[Sl 1] Slodowy, P.: Einige Bemerkungen zur Entfaltung symmetrischer Funktionen, Math. Zeitschrift 158, 157 - 170 (1978)

[Sl 2] Slodowy, P.: Four lectures on simple groups and singularities, Communications of the Mathematical Institute, Rijksuniversiteit Utrecht, Vol. 11, 1980

[Sp 1] Springer, T. A.: Some arithmetical results on semisimple Lie algebras, Publ. Math. I.H.E.S. 30, 115 - 141 (1966)

[Sp 2] Springer, T. A.: The unipotent variety of a semisimple group, Proc. of the Bombay Colloqu. in Algebraic Geometry, ed. S. Abhyankar, London, Oxford Univ. Press, 1969, 373 - 391

[Sp 3] Springer, T. A.: Invariant theory, Lecture Notes in Math. 585, Springer, Berlin-Heidelberg-New York, 1977

[S-S] Springer, T. A., Steinberg, R.: Conjugacy classes, in: Borel et alii: Seminar on algebraic groups and related finite groups, Lecture Notes in Math. 131, Springer, Berlin-Heidelberg-New York, 1970

[St 0] Steinberg, R.: Automorphisms of classical Lie algebras, Pacific J. of Math. 11, 1119 - 1129 (1961)

[St 1] Steinberg, R.: Regular elements of semisimple algebraic groups, Publ. Math. I.H.E.S. 25, 49 - 80 (1965)

[St 2] Steinberg, R.: Conjugacy classes in algebraic groups, Lecture Notes in Math. 366, Springer, Berlin-Heidelberg-New York, 1974

[St 3] Steinberg, R.: Torsion in reductive groups, Advances in Math. 15, 63 - 92 (1975)

[St 4] Steinberg, R.: On the desingularization of the unipotent variety, Inventiones math. 36, 209 - 224 (1976)

[Te] Teissier, B.: Cycles évanescents, sections planes et conditions de Whitney, Asterisque 7 et 8, 285 - 362 (1973)

[Th] Thom, R.: Stabilité structurelle et morphogénèse, Benjamin, Reading, Massachusetts, 1972

[Ti] Tits, J.: Classification of algebraic semisimple groups, in: Algebraic Groups and Discontinuous Subgroups, Proc. Symp. Pure Math. IX, ed. Borel, A., Mostow, G. D., A.M.S., 1966, 33 - 62

[Tj 1] Tjurina, G. N.: Locally semiuniversal flat deformations of isolated singularities of complex spaces, Math. USSR Izvestija, Vol. 3, No. 5, 967 - 999 (1969)

[Tj 2] Tjurina, G. N.: Resolutions of singularities of flat deformations of rational double points, Functional Anal. Appl. 4, 1, 68 - 73 (1970)

[V] Varadarajan, V. S.: On the ring of invariant polynomials on a semisimple Lie algebra, Amer. J. Math. 90, 308 - 317 (1968)

[Ve] Veldkamp, F. D.: The center of the universal enveloping algebra of a Lie algebra in characteristic p, Ann. Scient. Ec. Norm. Sup. 5, 217 - 240 (1972)

[W] Wahl, J.: Simultaneous resolution of rational singularities, Compositio Math. 38, 43 - 54 (1979)

Subject Index

Vol. 640: J. L. Dupont, Curvature and Characteristic Classes. X, 175 pages. 1978.

Vol. 641: Séminaire d'Algèbre Paul Dubreil, Proceedings Paris 1976–1977. Edité par M. P. Malliavin. IV, 367 pages. 1978.

Vol. 642: Theory and Applications of Graphs, Proceedings, Michigan 1976. Edited by Y. Alavi and D. R. Lick. XIV, 635 pages. 1978.

Vol. 643: M. Davis, Multiaxial Actions on Manifolds. VI, 141 pages. 1978.

Vol. 644: Vector Space Measures and Applications I, Proceedings 1977. Edited by R. M. Aron and S. Dineen. VIII, 451 pages. 1978.

Vol. 645: Vector Space Measures and Applications II, Proceedings 1977. Edited by R. M. Aron and S. Dineen. VIII, 218 pages. 1978.

Vol. 646: O. Tammi, Extremum Problems for Bounded Univalent Functions. VIII, 313 pages. 1978.

Vol. 647: L. J. Ratliff, Jr., Chain Conjectures in Ring Theory. VIII, 133 pages. 1978.

Vol. 648: Nonlinear Partial Differential Equations and Applications, Proceedings, Indiana 1976–1977. Edited by J. M. Chadam. VI, 206 pages. 1978.

Vol. 649: Séminaire de Probabilités XII, Proceedings, Strasbourg, 1976–1977. Edité par C. Dellacherie, P. A. Meyer et M. Weil. VIII, 805 pages. 1978.

Vol. 650: C*-Algebras and Applications to Physics. Proceedings 1977. Edited by H. Araki and R. V. Kadison. V, 192 pages. 1978.

Vol. 651: P. W. Michor, Functors and Categories of Banach Spaces. VI, 99 pages. 1978.

Vol. 652: Differential Topology, Foliations and Gelfand-Fuks-Cohomology, Proceedings 1976. Edited by P. A. Schweitzer. XIV, 252 pages. 1978.

Vol. 653: Locally Interacting Systems and Their Application in Biology. Proceedings, 1976. Edited by R. L. Dobrushin, V. I. Kryukov and A. L. Toom. XI, 202 pages. 1978.

Vol. 654: J. P. Buhler, Icosahedral Golois Representations. III, 143 pages. 1978.

Vol. 655: R. Baeza, Quadratic Forms Over Semilocal Rings. VI, 199 pages. 1978.

Vol. 656: Probability Theory on Vector Spaces. Proceedings, 1977. Edited by A. Weron. VIII, 274 pages. 1978.

Vol. 657: Geometric Applications of Homotopy Theory I, Proceedings 1977. Edited by M. G. Barratt and M. E. Mahowald. VIII, 459 pages. 1978.

Vol. 658: Geometric Applications of Homotopy Theory II, Proceedings 1977. Edited by M. G. Barratt and M. E. Mahowald. VIII, 487 pages. 1978.

Vol. 659: Bruckner, Differentiation of Real Functions. X, 247 pages. 1978.

Vol. 660: Equations aux Dérivée Partielles. Proceedings, 1977. Edité par Pham The Lai. VI, 216 pages. 1978.

Vol. 661: P. T. Johnstone, R. Paré, R. D. Rosebrugh, D. Schumacher, R. J. Wood, and G. C. Wraith, Indexed Categories and Their Applications. VII, 260 pages. 1978.

Vol. 662: Akin, The Metric Theory of Banach Manifolds. XIX, 306 pages. 1978.

Vol. 663: J. F. Berglund, H. D. Junghenn, P. Milnes, Compact Right Topological Semigroups and Generalizations of Almost Periodicity. X, 243 pages. 1978.

Vol. 664: Algebraic and Geometric Topology, Proceedings, 1977. Edited by K. C. Millett. XI, 240 pages. 1978.

Vol. 665: Journées d'Analyse Non Linéaire. Proceedings, 1977. Edité par P. Bénilan et J. Robert. VIII, 256 pages. 1978.

Vol. 666: B. Beauzamy, Espaces d'Interpolation Réels: Topologie et Géometrie. X, 104 pages. 1978.

Vol. 667: J. Gilewicz, Approximants de Padé. XIV, 511 pages. 1978.

Vol. 668: The Structure of Attractors in Dynamical Systems. Proceedings, 1977. Edited by J. C. Martin, N. G. Markley and W. Perrizo. VI, 264 pages. 1978.

Vol. 669: Higher Set Theory. Proceedings, 1977. Edited by G. H. Müller and D. S. Scott. XII, 476 pages. 1978.

Vol. 670: Fonctions de Plusieurs Variables Complexes III, Proceedings, 1977. Edité par F. Norguet. XII, 394 pages. 1978.

Vol. 671: R. T. Smythe and J. C. Wierman, First-Passage Perculation on the Square Lattice. VIII, 196 pages. 1978.

Vol. 672: R. L. Taylor, Stochastic Convergence of Weighted Sums of Random Elements in Linear Spaces. VII, 216 pages. 1978.

Vol. 673: Algebraic Topology, Proceedings 1977. Edited by P. Hoffman, R. Piccinini and D. Sjerve. VI, 278 pages. 1978.

Vol. 674: Z. Fiedorowicz and S. Priddy, Homology of Classical Groups Over Finite Fields and Their Associated Infinite Loop Spaces. VI, 434 pages. 1978.

Vol. 675: J. Galambos and S. Kotz, Characterizations of Probability Distributions. VIII, 169 pages. 1978.

Vol. 676: Differential Geometrical Methods in Mathematical Physics II, Proceedings, 1977. Edited by K. Bleuler, H. R. Petry and A. Reetz. VI, 626 pages. 1978.

Vol. 677: Séminaire Bourbaki, vol. 1976/77, Exposés 489–506. IV, 264 pages. 1978.

Vol. 678: D. Dacunha-Castelle, H. Heyer et B. Roynette. Ecole d'Eté de Probabilités de Saint-Flour. VII-1977. Edité par P. L. Hennequin. IX, 379 pages. 1978.

Vol. 679: Numerical Treatment of Differential Equations in Applications, Proceedings, 1977. Edited by R. Ansorge and W. Törnig. IX, 163 pages. 1978.

Vol. 680: Mathematical Control Theory, Proceedings, 1977. Edited by W. A. Coppel. IX, 257 pages. 1978.

Vol. 681: Séminaire de Théorie du Potentiel Paris, No. 3, Directeurs: M. Brelot, G. Choquet et J. Deny. Rédacteurs: F. Hirsch et G. Mokobodzki. VII, 294 pages. 1978.

Vol. 682: G. D. James, The Representation Theory of the Symmetric Groups. V, 156 pages. 1978.

Vol. 683: Variétés Analytiques Compactes, Proceedings, 1977. Edité par Y. Hervier et A. Hirschowitz. V, 248 pages. 1978.

Vol. 684: E. E. Rosinger, Distributions and Nonlinear Partial Differential Equations. XI, 146 pages. 1978.

Vol. 685: Knot Theory, Proceedings, 1977. Edited by J. C. Hausmann. VII, 311 pages. 1978.

Vol. 686: Combinatorial Mathematics, Proceedings, 1977. Edited by D. A. Holton and J. Seberry. IX, 353 pages. 1978.

Vol. 687: Algebraic Geometry, Proceedings, 1977. Edited by L. D. Olson. V, 244 pages. 1978.

Vol. 688: J. Dydak and J. Segal, Shape Theory. VI, 150 pages. 1978.

Vol. 689: Cabal Seminar 76–77, Proceedings, 1976–77. Edited by A.S. Kechris and Y. N. Moschovakis. V, 282 pages. 1978.

Vol. 690: W. J. J. Rey, Robust Statistical Methods. VI, 128 pages. 1978.

Vol. 691: G. Viennot, Algèbres de Lie Libres et Monoïdes Libres. III, 124 pages. 1978.

Vol. 692: T. Husain and S. M. Khaleelulla, Barrelledness in Topological and Ordered Vector Spaces. IX, 258 pages. 1978.

Vol. 693: Hilbert Space Operators, Proceedings, 1977. Edited by J. M. Bachar Jr. and D. W. Hadwin. VIII, 184 pages. 1978.

Vol. 694: Séminaire Pierre Lelong – Henri Skoda (Analyse) Année 1976/77. VII, 334 pages. 1978.

Vol. 695: Measure Theory Applications to Stochastic Analysis, Proceedings, 1977. Edited by G. Kallianpur and D. Kölzow. XII, 261 pages. 1978.

Vol. 696: P. J. Feinsilver, Special Functions, Probability Semigroups, and Hamiltonian Flows. VI, 112 pages. 1978.

Vol. 697: Topics in Algebra, Proceedings, 1978. Edited by M. F. Newman. XI, 229 pages. 1978.

Vol. 698: E. Grosswald, Bessel Polynomials. XIV, 182 pages. 1978.

Vol. 699: R. E. Greene and H.-H. Wu, Function Theory on Manifolds Which Possess a Pole. III, 215 pages. 1979.